Amy Stewart completed her PhD at the University of Newcastle upon Tyne and has worked for the Forestry Commission of Great Britain, participating in projects for the delivery of social, stewardship and conservation targets at the local level, and engaging with national and international policy issues. Her research interests focus on sustainable development, especially globally, and she has co-authored an article on international multi-stakeholder partnerships for sustainable development.

Tim Gray is Emeritus Professor of Political Thought at Newcastle University. His research interests centre on environmental politics, especially in the field of natural resource management, such as fisheries and water, and his recent co-authored publications include papers on participation in fisheries governance, the discard problem, fisheries-dependent communities, the Common Fisheries Policy, and offshore wind farms. He is the author of 'Freedom' (Issues in Political Theory, 1991) and editor of 'UK Environmental Policy in the 1990s' (1995), 'The Politics of Fishing' (1998) and 'Participation in Fisheries Governance' (2005).

INTERNATIONAL LIBRARY OF AFRICAN STUDIES

4. *The Nigerian Military and the State*
Jimmy Peters
978 1 85043 874 8

5. *Adjusting Society: The World Bank, the IMF and Ghana*
Lynne Brydon & Karen Legge
978 1 86064 000 1

6. *The Horn of Africa: Politics and International Relations*
Peter Woodward
978 1 85043 741 3

7. *State of Emergency: Nyasaland 1959*
Colin Baker
978 1 86064 068 1

8. *From Colonization to Democracy: A New Historical Geography of South Africa* – Alan Lester
978 1 86064 091 9

10. *South Africa in Transition: The Misunderstood Miracle*
Adrian Guelke
978 1 86064 343 9

11. *Security and the State in Southern Africa* – Agostinho Zacarias
978 1 86064 328 6

12. *Coming to Terms: Zimbabwe in the International Arena* – Richard Schwartz
978 1 86064 647 8

13. *Britain, Kenya and the Cold War: Imperial Defence, Colonial Security and Decolonisation* – David Percox
978 1 85043 460 3

14. *Apartheid South Africa and African States: From Pariah to Middle Power, 1961–1994* – Roger Pfister
978 1 85043 625 6

15. *A History of the Left in South Africa: Writings of Baruch Hirson*
Baruch Hirson
978 1 85043 454 2

16. *An African Trading Empire: The Story of Susman Brothers & Wulfsohn, 1901–2005*
Hugh Macmillan
978 1 85043 853 3

17. *The Place of Tears: The Novel and Politics in Modern Zimbabwe*
Ranka Primorac
978 1 84511 120 5

18. *Mineworkers in Zambia: Labour and Political Change in Post-Colonial Africa*
Miles Larmer
978 1 84511 299 8

19. *Reconstructing the Nation in Africa: The Politics of Nationalism in Ghana*
Michael Amoah
978 1 84511 259 2

20. *Revolt and Protest: Student Politics and Activism in Sub-Saharan Africa*
Leo Zeilig
978 1 84511 476 3

21. *The Other Way Around* – Frank Verdon
978 1 86064 030 8

22. *The Politics of Neoliberal Democracy in Africa: State and Civil Society in Nigeria*
Usman A. Tar
978 1 84511 656 9

23. *Afrikaners in the New South Africa: Identity Politics in a Globalised Economy*
Rebecca Davies
978 1 84511 785 6

24. *The Politics of Water in Africa: The European Union's Role in Development Aid Partnership* – Chris Rowan
978 1 84511 685 9

25. *African Women and Apartheid: Migration and Settlement in Urban South Africa* – Rebekah Lee
978 1 84511 819 8

26. *Islam's Perfect Stranger: The Life of Mahmud Muhammad Taha, Muslim Reformer of Sudan* – Edward Thomas
978 1 84885 0 040

27. *The Governance of Water and Sanitation in Africa: Achieving Sustainable Development Through Partnerships*
Tim Gray and Amy Stewart
978 1 84885 0 279

THE GOVERNANCE OF WATER AND SANITATION IN AFRICA

Achieving Sustainable Development through Partnerships

AMY STEWART AND TIM GRAY

TAURIS ACADEMIC STUDIES
an imprint of
I.B.Tauris Publishers
LONDON • NEW YORK

Published in 2009 by Tauris Academic Studies,
An imprint of I.B.Tauris & Co Ltd
6 Salem Road, London W2 4BU
175 Fifth Avenue, New York NY 10010
www.ibtauris.com

Distributed in the United States and Canada Exclusively by Palgrave Macmillan
175 Fifth Avenue, New York NY 10010

International Library of African Studies 27

ISBN 978 1 84885 027 9

A full CIP record for this book is available from the British Library
A full CIP record for this book is available from the Library of Congress

Library of Congress catalog card: available

Printed and bound in India by Thomson Press (India)
Camera-ready copy edited and supplied by the author

This book is dedicated to Amy's parents, sister and brothers, for their unwavering belief in this project and for their unconditional support

CONTENTS

ACKNOWLEDGEMENTS

This book arose out of Amy Stewart's PhD thesis supervised by Tim Gray at Newcastle University 2003–7, funded by an ESRC Studentship. We acknowledge with gratitude the support provided by both institutions. We also thank the colleagues and friends who supported us during the writing of the thesis and the book, especially Sarah and Lucy.

ACRONYMS

ACP-EU	African, Caribbean and Pacific – European Union
ADPs	Area Development Programmes [World Vision]
A-EUWI	African – European Union Water Initiative
AMCOW	African Ministers' Council on Water
AMCOW-EXCO	African Ministers' Council on Water Executive Committee
AMCOW-TAC	African Ministers' Council on Water – Technical Advisory Committee
ANEW	African Civil Society Network on Water and Sanitation
APWO	Association of Private Water Operators [Uganda]
ARD	Associates in Rural Development
CBO	Community-Based Organization
CCC	Chlorine Chemistry Council
CCCM	Customer Care Charter Model
CGG	Commission on Global Governance
CIARIS	Centre for Informatic Apprenticeship and Resources in Social Exclusion
CMA	Catchment Management Agency
CMI	Chr. Michelsen Institute
CNHF	Conrad N. Hilton Foundation
COMPASS	Community Partnerships for Sustainable Resource Management in Malawi
CSD	Commission on Sustainable Development [UN]
CSO	Civil Society Organization

DANIDA	Danish International Development Agency
DCI	Development Cooperation Ireland
DEFRA	Department for Environment, Food and Rural Affairs [UK]
DFID	Department for International Development [UK]
DG	Directorate-General [EU Commission]
DPLG	Department for Provincial and Local Government [South Africa]
DRI	Desert Research Institute
DTF	Devolution Trust Funds
DWAF	Department of Water Affairs and Forestry [South Africa]
DWA	Department of Water Affairs [Zambia]
DWD	Directorate of Water Development [Uganda]
EA	Environment Agency [UK]
EDF	European Development Fund
EIB	European Investment Bank
EU	European Union
EUWI	European Union Water Initiative
GBS	General Budget Support
GIS	Geographic Information System
GTZ	German Technical Co-operation
HIP	Harmonization in Practice
HKW	Helen Keller Worldwide
IDSUH	Institute of Development Studies – University of Helsinki
IDSRU	Department of International Development Studies – Roskilde University Centre
IGO	International Governmental Organization
IO	International Organization
ITI	International Trachoma Initiative
IWRM	Integrated Water Resource Management
JASZ	Joint Assistance Strategy for Zambia

JDSD	Johannesburg Declaration on Sustainable Development
JICA	Japanese International Cooperation Agency
JMP	Joint Monitoring Programme [WHO/UNICEF]
JPOI	Johannesburg Plan of Implementation
MDG	Millennium Development Goal
MEWD	Ministry of Energy and Water Development [Zambia]
MLGH	Ministry of Local Government and Housing [Zambia]
MoFNP	Ministry of Finance and National Planning [Zambia]
MoU	Memorandum of Understanding
MS	Member States [EU]
MSF	Multi-Stakeholder Forum
NDP	National Development Plan
NEEDS	National Economic Empowerment and Development Strategy [South Africa]
NEPAD	New Partnership for Africa's Development
NGO	Non-Governmental Organization
NGDO	Non-Governmental Development Organization
NWASCO	National Water Supply and Sanitation Council [Zambia]
ODA	Official Development Assistance
OFWAT	Water Services Regulation Authority (formally Office of Water Service) [UK]
PAWS	Partners for Water and Sanitation
PPP	Public-Private Partnership
PRSP	Poverty Reduction Strategy Paper
RWSS	Rural Water Supply and Sanitation
SADC	Southern African Development Community
SALGA	South African Local Government Association

SANGOCO	South African NGO Coalition
SG	Steering Group
SWAP	Sector Wide Approach
ToR	Terms of Reference
TNC	Transnational Corporation
TTP	Type Two Partnership
UNCED	United Nations Conference on Environment and Development [Rio Earth Summit 1992]
UNCSD	United Nations Commission on Sustainable Development
UNDESA	United Nations Department of Economic and Social Affairs
UNDP	United Nations Development Programme
UNDSD	United Nations Division for Sustainable Development
UNESC	United Nations Economic and Social Council
UNICEF	United Nations Children's Fund
UNMPTF-WS	United Nations Millennium Project Task Force on Water and Sanitation
USAID	United States Agency for International Development
UWASNET	Ugandan Water and Sanitation NGO Network
UWSS	Urban Water Supply and Sanitation
WASAZA	Water and Sanitation Association of Zambia
WAWI	West Africa Water Initiative
WCA	Water Consumer Association
WCC	World Chlorine Council
WF	Water Facility [EU]
WG	Working Group
WHO	World Health Organization
WRAP	Water Resources Action Programme [Zambia]
WSP	Water and Sanitation Programme
WSS	Water Supply and Sanitation
WSSD	World Summit on Sustainable Development
WWF	World Wide Fund for Nature
ZDM	Zululand District Municipality

EXECUTIVE SUMMARY

The outcomes of the UN World Summit on Sustainable Development (WSSD), held in Johannesburg in 2002, were noteworthy, not for the traditional, formal governmentally-agreed commitments (Type One agreements) but for the advent of voluntary, multi-stakeholder partnerships (Type Two Partnerships or TTPs) aimed at complementing governmental efforts to achieve goals outlined in international sustainable development agreements. At the WSSD, governments agreed to halve the number of people worldwide who lacked access to safe water and sanitation by 2015. This book examines three of the relatively few TTPs which focused on helping to achieve the drinking water and sanitation goals in Africa, where water issues were most acute: the EU Water Initiative (EUWI); Partners for Water and Sanitation (PAWS); and the West Africa Water Initiative (WAWI). By undertaking a comparative analysis of these case studies, our aim is to evaluate the worth of TTPs as governance mechanisms, in terms of both processes and outputs. The conceptual framework used to underscore this analysis has two elements – the concept of sustainable development, and the concept of partnership. The methodology we employed involved fieldwork in Europe, the USA, South Africa, Zambia and Ghana, and relied heavily on semi-structured interviews undertaken with key players associated with these partnerships, as well as documentary analysis. We argue that although the three cases studied all displayed deficiencies (in terms of both their partnership processes, and their outputs), the TTP framework is a valuable initiative, with the potential to significantly improve the delivery of international sustainable development objectives, (including water and sanitation provision) in the long-term.

1

INTRODUCTION

The Problem and the 'Type Two' Solution

It is estimated that in the world today just over one billion people have no access to a safe water supply, and more than double that figure lack adequate sanitation, the majority of these individuals living in the developing world. 'The 1.8 million child deaths each year related to unclean water and poor sanitation dwarf the casualties associated with violent conflict. No act of terrorism generates economic devastation on the scale of the crisis in water and sanitation' (UNDP 2006, 3). Although water is a finite resource, and 'Essentially there is no more fresh water on the planet today than there was two thousand years ago, when the global population was less than three percent of what it is now' (Knighton 2002, 13), the problem is that 'The world water crisis is a crisis of governance – not one of scarcity. At the global scale, there is enough water to provide "water security" for all, but only if we change the way we manage and develop it' (HRH the Prince of Orange 2002, 2–3). This is a view with which the United Nations Development Programme (UNDP 2006, v) concur, stating that, 'the roots of the crisis in water can be traced to poverty, inequality and unequal power relationships, as well as flawed water management policies that exacerbate scarcity.'

In 2002, the World Summit on Sustainable Development (WSSD) took place in Johannesburg, South Africa, and in the run-up to this conference, the UN Secretary General, Kofi Annan, identified five priority issues to be addressed: water, energy, health, agriculture, and biodiversity. Water was one of the few issues upon which a clear objective was agreed at the summit, whereby governments reiterated their support for the Millennium Development Goal (MDG) to halve the number of people lacking access to safe water by 2015, with the baseline set as 1990. They also added a further commitment within the same timeframe, and with the same baseline: to halve the number of people worldwide lacking basic sanitation. Water is not only a basic human need, but is central to poverty eradication and sustainable development, and 'is intimately linked' to all of the priority issues addressed at the WSSD (Mwanza 2003, 95). Indeed, 'Without progress on water [and sanitation], reaching other Millennium Development Goals . . . will be difficult if not impossible' (Mwanza 2003, 95), since 'it is crucial to all forms of social and economic development' (Berntell 2005, 3). In other words, 'Clean water and sanitation can make or break human development. They are fundamental to what people can do or what they can become – to their capabilities' (UNDP 2006, 27).

Although the 2004 WHO/UNICEF Joint Monitoring Programme for Water Supply and Sanitation report assessing the mid-term progress towards meeting the water and sanitation MDG targets, found that 'the world is on track to meet the MDGs target for drinking water,' this 'news is tempered . . . by slow progress in sub-Saharan Africa and stalled action on sanitation in most developing regions' (WHO/UNICEF JMP 2004, 6). Indeed, the report found that in terms of safe drinking water, 'In sub-Saharan Africa, 42 per cent of the population is still unserved' (WHO/UNICEF JMP 2004, 8), and in terms of sanitation, 'In sub-Saharan Africa . . . coverage is a mere 36 per cent' (WHO/UNICEF JMP 2004, 12). Furthermore, as the UN Millennium Project Task Force on Water and Sanitation (an independent advisory body commissioned by the UN Secretary General and the administrator of the UN Development Programme to help the international community reach the water and sanitation MDGs) pointed out, 'Significant progress in China and India alone, for instance, could achieve the global target – without there being any progress at all in Sub-Saharan Africa,' yet it was in Africa that many of the people in greatest need lived (UNMPTF-WS 2005, 21).

Water and sanitation problems are not new and nor are international attempts to tackle them, so the intergovernmentally-agreed target-setting seen at the WSSD was not particularly groundbreaking. However, 'for the first time, outcome documents were not the sole product' of an international sustainable development summit (UNDESA 2002a). As Zadek (quoted in Martens 2007, 16) observed, 'The Johannesburg Summit was more than anything about partnerships'. Alongside the formal, intergovernmental, or 'Type One' agreements, was the advent of a new voluntary framework for partnerships, called 'Partnerships for Sustainable Development' or 'Type Two Partnerships' (TTPs), with the intention that these partnerships should 'contribute to and reinforce the implementation of the outcomes of the intergovernmental negotiations of the WSSD,' and should complement governmental commitments, though not replace them (Kara & Quarless 2002, 1). Norris (2005, 212) explained that while TTPs were 'informal, voluntary commitments, generally between a combination of governments, businesses, NGOs [non-governmental organizations] and IGOs [international governmental organizations],' Type One agreements were 'specific commitments between governments.' Pallemaerts (2003, 283) reported that 'the United Nations' interest in partnerships can be traced back to the Millennium Declaration of September 2000, in which heads of State and Government resolved "to develop strong partnerships with the private sector and with civil society organizations in pursuit of development and poverty eradication".' In the following two years, the General Assembly adopted two resolutions on partnerships as a means of contributing to the achievement of UN programmes and goals and 'The WSSD became the first large-scale testing ground for the new partnership approach' (Pallemaerts 2003, 283).

An important part of the thinking behind the TTP initiative was frustration at the failure of international law to deliver commitments made by nation states in the past. As Bruch & Pendergrass (2003, 855; 856) noted, critics argued that 'the commitments made at Rio had not been fulfilled, and the last thing the international community needs is more international law that will fail to be implemented': a far more promising approach than further coercion was the voluntary principle – 'voluntary partnerships to implement the existing commitments'. This is not to say that voluntarism was to replace regulation; rather it was to complement it (Hale & Mauzerall 2004, 221).

At the fourth preparatory meeting for the WSSD in Bali, in May 2002, guiding principles were drawn up for this new framework, known as the Bali Guiding Principles, and these were further developed at the first follow-up meeting to the WSSD, the 11th session of the Commission on Sustainable Development (CSD), in May 2003. At Bali, it was affirmed that partnerships should 'have a multi-stakeholder approach . . . arranged among any combination of partners, including governments, regional groups, local authorities, non-governmental actors, international institutions and private sector partners' (Kara & Quarless 2002, 2). Non-governmental stakeholders were, for the first time, officially recognized as having a role to play in the practical delivery of international sustainable development agreements.

In the Johannesburg Plan of Implementation (JPOI) (a formal, governmentally-agreed outcome document of the WSSD), the UN CSD Secretariat was designated as the focal point for discussion on TTPs, and partnerships were encouraged to register with the Secretariat, and report to the Secretariat on their progress, at least on a biennial basis.

Over 200 partnerships were launched during the WSSD process (UNESC 2004, 3), and by January 2008, a total of 334 such partnerships had been registered (UNDSD 2008a). Of these, 77 reported a primary focus on water (the largest for any thematic grouping), and 58 reported a secondary focus on water (although most partnerships had multiple primary and secondary themes) (UNDSD 2008b). However, only a small number of these concentrated specifically, and directly, on drinking water and sanitation provision, rather than on broader issues of water management, and an even smaller number concentrated on these issues in a sub-Saharan African context, where 'Water problems are most acute' (EU 2002, 1). This book directs attention to three of the relatively few TTPs focusing their efforts on drinking water and sanitation provision in sub-Saharan Africa: European Union Water Initiative (EUWI); Partners for Water and Sanitation (PAWS); and West Africa Water Initiative (WAWI).

The hypothesis that we test in this book is that the TTP framework could help to overcome some of the deficiencies of previous international, bilateral and multi-lateral development efforts because of its participatory, multi-stakeholder partnership approach, since this invests it with the potential to create more holistic and innovative understandings of problems to be addressed and solutions to these problems. The TTP framework promises to avoid duplication of efforts; to foster a sense of indigenous ownership over policies and their implementation; to empower beneficiaries; to mobilize new resources (not just financial, but also social capital of knowledge, skills, and expertise) beyond those of governments; and ultimately to achieve more than could be done by any single actor working alone. By undertaking a comparative analysis of three case studies, the aim of this book is to evaluate the worth of the

TTP framework as a governance mechanism, in terms of both processes and outputs, and assess how far the theoretical potential of the TTP framework has been realized in practice. The analysis will be more heavily weighted towards assessing procedural and structural issues than outcomes delivered, because, first, the partnerships examined are too new to be judged on whether they have yet delivered on their output goals; and second, if the correct processes of partnership are not in place, then the target aspirations of the TTP framework cannot be fully met.

Our central argument is that the potential benefits of the TTP approach are not just a pipe dream – multi-stakeholder partnerships can play an important role in sustainable development governance – but the full potential of this approach has not yet been realized. One of the largest hurdles to the fulfilment of the potential of the TTP approach is the role of government. Governments must not entirely relinquish their responsibilities to other actors, but nor should they manipulate these other actors; a balancing act is required between governmental leadership and genuine stakeholder participation, whereby governments preserve a degree of sovereignty and take ultimate responsibility for service provision and international development, but engage with other stakeholders to share responsibility for the management of public affairs and the implementation of sustainable development. None of the case studies examined is without its deficiencies – inequitable relationships; weak modes of partnership; and insufficient engagement with certain stakeholders (both governmental and non-governmental) have been evident in all three cases. However, evidence suggests that TTPs can achieve much more than either solo efforts by individuals or organizations working alone, or bilateral initiatives by governments, and the partnerships investigated here are beginning to demonstrate what operating in partnership can accomplish. Nonetheless, more needs to be done to monitor and evaluate TTPs to ensure that they are accountable, that their legitimacy remains intact, and that lessons can be learned about which partnership modes work, and which do not.

In the next section of this chapter, we review the relevant literature, situating the book in the context of contemporary debate. In chapter two, we develop our conceptual framework, built upon the two seminal concepts of sustainable development (with its associated concepts of participation and governance), and partnership (with its associated concept of stakeholder). The following three chapters each focus on an individual partnership: chapter three on EUWI; chapter four on PAWS; and chapter five on WAWI. Chapter six draws on the findings of these empirical chapters, comparing and contrasting them and relating discussion back to the partnership framework established in chapter two. Finally, chapter seven closes the study with an overview of the findings of the book and a discussion of the wider implications of these findings in relation to the sustainable development framework established in chapter two.

Relevant Literature

Critical Literature on the Type Two Partnership Framework

While several studies investigated multi-stakeholder partnerships (see chapter two), very little has been written about TTPs specifically, especially in the academic literature. In what does exist, there has often been expressed the hope that the TTP framework:

1) Would improve the implementation strategies for sustainable development because it brought together a wider range of capabilities and knowledge than previously had been the case, and would therefore deliver better informed decisions and solutions (Ayre & Callway 2005, 15; Calder 2002, 8; Eweje 2007; Hemmati & Whitfield 2003b, 4; Ivanova 2003, 13).
2) Would address the fact that sustainable development could not be tackled by any one actor working alone and required a sense of shared responsibility (Eweje 2007, 24; Hemmati & Whitfield 2003a, 3; Ivanova 2003, 11; Küpçü's 2005, 91; Müller-Kraenner 2003, 55; Witte *et al.* 2003, 61).
3) Would increase participation and therefore empower non-governmental stakeholders to take an active role in the implementation of policies which affected them (Hemmati & Whitfield 2003b, 4).
4) Would help to overcome the inequalities between the developed countries of the northern hemisphere, and developing countries in the south – moving the development agenda away from conditionality, towards co-operation and country-owned strategies (Calder 2002, 8).

Underlying most (although not all) these views there is a common thread – 'that partnerships *properly* conceived and managed – can in fact make a useful contribution to global environmental governance' (Witte & Streck 2003, 5). However, convergence on what constitutes "proper" is a contested matter, and writers have conveyed the following (sometimes conflicting) fears that the TTP framework:

1) Would be exploited by governments at the WSSD to reduce pressure on them to agree binding commitments and to relinquish their responsibilities for implementation (Ayre & Callway 2005, 23; Calder 2002, 8; Coates 2002; Friends of the Earth 2002; Gardiner 2002, 5; Hemmati & Whitfield 2003b, 3; Küpçü 2005, 95; Pallemaerts 2003, 282; Witte & Streck 2003, 3).
2) Would only be successful if governments were willing to set binding and ambitious targets because 'they have no legitimacy in their own right' (Witte *et al.* 2003, 70–71).
3) Would be inadequately linked to official intergovernmental processes and global governance processes, including the formal outcomes of the WSSD (Ayre & Callway 2005, 23; Gardiner 2002, 5; Ivanova 2003, 17; Witte *et al.* 2003, 61).
4) Would lack sufficient mechanisms and rules to guarantee the transparency and accountability of TTP initiatives and for them to be monitored effectively (Ayre & Callway 2005, 38; Friends of the Earth 2002; Hemmati & Whitfield 2003b, 5; Küpçü 2005, 95; Müller-Kraenner 2003, 57; Witte & Streck 2003, 4; Zadek 2003, 10).

However, few of these authors applied their analyses of the TTP framework to specific TTPs in any depth, confining their attention to largely theoretical conjecture. There is, therefore, a need for an empirical analysis of particular TTPs, in order to assess their limitations and drawbacks, as well as their potential for improving the way sustainable development is governed at the international, as well as local level. The UN published three reports on TTPs: the 2004, *Commission on Sustainable Development Twelfth Session. Partnerships for Sustainable Development – Report of the Secretary-General* (UNESC 2004); the 2005, *Commission on Sustainable Development Thirteenth Session: Partnerships for Sustainable Development – Update* (UNDESA 2005); and the 2006, *Commission on Sustainable Development Fourteenth Session: Partnerships for Sustainable Development*

– *Report of the Secretary-General* (UNESC 2006a). These reports were intended to review the 'contribution of partnerships to the implementation of the intergovernmentally agreed sustainable development goals and objectives' (UNESC 2004). However, their analysis was insubstantial and partial since it was limited to a synthesis of facts and figures, such as the number of partnerships as defined by sector; the geographic scope of registered partnerships; their time-frame; their membership and resource base; as well as implementation trends. While this analysis is useful and important, in order to fully appreciate the value of the TTP framework, a comprehensive assessment of individual TTPs is necessary.

This book sets out to provide such an assessment, by addressing all of the issues identified by the authors above, through first, clarifying the principles of partnership necessary for the positive potential of TTPs to be brought to fruition; second, looking at the implications of the TTP framework for governance for sustainable development; third, using empirical case studies to investigate how far individual TTPs were living up to the aspirations associated with them; and fourth, considering whether the concerns raised by critics of TTPs were well-founded in practice. Ivanova (2003, 17) pointed out that 'partnerships are [not] effective by definition . . . Their ultimate effectiveness will hinge upon their internal structure as well as their relationship to the broader forms of global governance.' This study has been conducted precisely to investigate the effectiveness of the internal structure and wider relationships of specific TTPs. We endorse the view that: 'partnerships must be evaluated to evolve' (Küpçü 2005, 96), and only through evaluation, such as that undertaken in the chapters that follow, will we begin to understand the contribution TTPs can potentially make towards the governance and implementation of sustainable development.

Documentary Literature on the Case Studies

The documentary literature on the three case study partnerships varies considerably in extent between EUWI, PAWS and WAWI. A wealth of documentation has been produced by the EUWI partnership and its members including: a lengthy launch document (EU 2002); a well-developed website (EUWI 2004d; EUWI 2006a); minutes and summaries of meetings held in both Africa and Europe (A-EUWI Zambia

2005a; A-EUWI Zambia 2005b; EUWI 2003a; EUWI 2003b; EUWI 2003c; EUWI 2004a; EUWI 2005a; EUWI 2005b; EUWI 2005c; EUWI 2005g; EUWI 2005h; EUWI Africa Working Group 2007; Schild 2004); codes of conduct, guidelines for engagement, and organizational framework documents (AMCOW 2006; EUWI 2005i; Walshe *et al.* 2004); country specific, status and country dialogue reports (EUWI 2004b; EUWI 2004e; EUWI Working Group on Water Supply and Sanitation in Africa 2005); discussion papers and presentations (EUWI 2005d; EUWI 2005f; Mbassi 2004); strategy reports, action plans, and work programmes (AMCOW-TAC & EU 2003; EUWI 2004c; EUWI 2005e; EUWI 2006b; EUWI Africa Working Group 2006a; EUWI Africa Working Group 2006b; EUWI Working Group on Water Supply and Sanitation in Africa 2004); and reviews of progress and lessons learned (A-EUWI 2005; EUWI 2007; EUWI Secretariat 2006; Mushauri 2006; WaterAid & Tearfund 2005). While some of this information is available in the public domain, the most useful literature for evaluation was only accessible via personal contacts and through membership of EUWI's Intranet.

The PAWS partnership also produced a significant amount of documentation, although much less than the EUWI since it operates on a smaller scale, including annual reports detailing work completed and plans for the following year (PAWS 2004b; 2005e; PAWS 2006j); terms of reference and rules for engagement (PAWS 2003g; PAWS 2005a); newsletters keeping interested parties up to speed with the latest developments (PAWS 2005c; PAWS 2005d; PAWS 2006d; PAWS 2006i; PAWS 2006k; PAWS 2007); and the partnership website which was regularly updated (PAWS 2003f; PAWS 2006a; PAWS 2006b; PAWS 2006e; PAWS 2006f; PAWS 2006g; PAWS 2006h; PAWS 2006i). The PAWS Secretariat also provided us with minutes and notes from partnership meetings (PAWS 2003e; PAWS 2004c; PAWS 2004d; PAWS 2005b); scoping study reports (PAWS 2003a; PAWS 2003b; PAWS 2003c; PAWS 2003d); and partnership visit reports (PAWS 2004e; PAWS 2004f), none of which are publicly available.

Documentation available for public scrutiny of WAWI was much more scant. Short documents existed on various WAWI members' websites, the initiative's own website, and the CSD's TTP database, which detailed the basic facts about WAWI's inception; its funding; membership; partner roles; partnership focus; timetables and targets (Chlorine Chemistry

Council 2004; CNHF 2002; CNHF 2003; CNHF 2004a; CNHF 2004b; CNHF 2004c; CNHF 2004d; Murray 2002; USAID 2002a; USAID 2002b ; USAID Water Team 2004a; USAID Water Team 2004b; Water for the Poor 2002a; Water for the Poor 2002b; Water for the Poor 2002c; World Vision 2002). Although the WAWI website was rarely updated, it did detail partnership objectives, and progress on these objectives in each country (Water for the Poor 2005a; Water for the Poor 2005b; Water for the Poor 2005c; Water for the Poor 2005d). Under its 'Publications' banner, reports, surveys, frameworks, assessments, and evaluations were also included. However, of the 18 items listed, only three could be viewed via the website, the rest having restricted or limited circulation. Regrettably, unlike with EUWI and PAWS, we could not obtain access to any of WAWI's documents that were not in the public domain.

Interview Data

Part of the uniqueness of this study comes, first, from access to internal documents not readily available in the public domain in the cases of EUWI and PAWS; and second, from interviews with partnership members and others, which enabled us to contextualize the documentary data, and assess its implications, for each partnership. Amy Stewart conducted 64 interviews: 31 interviewees were connected to EUWI, 21 to PAWS, four to both PAWS and EUWI, and eight to WAWI. The number of interviewees relating to each partnership was related to the size of the partnership, i.e. the number of members it had, as well as ease of access to respondents. Since EUWI is a massive partnership in membership terms compared with PAWS and WAWI, it was necessary to interview more people from EUWI in order to gain a representative overview. PAWS is larger than WAWI in membership terms and this was again reflected in the number of participants interviewed.

The choice of interviewees was dictated partly by access, and partly by the need to ensure that the views of a broad range of respondents from each partnership were incorporated into the study, and that representatives from all the major types of stakeholder (i.e. civil society, the private sector, and governments) in each partnership were included. Snowballing was also used, asking respondents who they thought should be interviewed. Some of the interviewees were happy to be interviewed for this study, but negotiating the interviews of many others was not easy.

Nearly all of the interviews were recorded, with the permission of the interviewees, and agreement was reached beforehand on how the material would be used. The interviews were transcribed, and the transcriptions were sent to the interviewees, giving them the opportunity to amend or correct them as they saw fit, and confirmation was obtained about how they wished the transcripts to be used. As a result, quotes from 29 of the interviews have been anonymized at the request of the respondents, and quotes from the remaining 35 respondents are referenced by name. However, the majority of respondents wished it to be made clear that the opinions they expressed were personal to themselves, and did not necessarily represent the views of the organizations with which they were associated.

Let us now turn to our conceptual framework in sustainable development and partnerships.

2

CONCEPTUAL FRAMEWORK

In this chapter, we explain the conceptual framework within which this study is embedded. There are two concepts which make up this framework – sustainable development and partnership. The concept of sustainable development supplies the overall organising idea out of which TTPs arose. The concept of partnership constitutes the working principle which guides the practice of TTPs. In other words, sustainable development provides the end, while partnership provides the means.

The Concept of Sustainable Development

Origin of the Concept of Sustainable Development

The concept of sustainable development is now well established as a dominant and pervasive policy discourse at the international, regional, national, and local levels (Bruyninckx 2006, 265). The foundations of the concept began in the 1970s when orthodox notions of growth and development came under question. Until this point, development was generally conceived as being a linear process moving from traditional societies to those based on mass-consumption. However, as Harris and Goodwin (2001, xxxiii) argued, a tension had emerged 'between the promotion of economic growth and the equitable provision of basic needs,' and there was a growing acknowledgement that 'From an ecological perspective, both human population and total resource demand must be limited in scale, and the integrity of ecosystems and diversity of species must be maintained.'

However, it was not until 1987 that the sustainable development concept gained world-wide status and significance, with the publication by the World Commission on Environment and Development (WCED), of *Our Common Future*, or as it has come to be known, the Brundtland Report, which represented 'a global agenda for change', incorporating both ecological and socio-economic objectives (WCED 1987, ix). Here the concept of sustainable development was defined as development that 'meets the needs of the present without compromising the ability of future generations to meet their own needs' (WCED 1987, 8). Brundtland claimed that meeting the essential needs of the world's poor – clean water supply and sanitation included – would require 'not only a new era of economic growth for nations in which the majority are poor, but an assurance that those poor get their fair share of the resources required to sustain that growth' (WCED 1987, 8). The Brundtland conception of sustainable development sought to integrate environmental and developmental concerns, 'thus breaking the perception that environmental protection can only be achieved *at the expense of* economic development' and instead focusing on 'how to achieve environmentally sustainable forms of development' (Baker *et al.* 1997, 3).

However, many critics have pointed to ambiguities in the Brundtland Report's definition of sustainable development, which allowed it to be evoked by those endeavouring to meet entirely opposing goals. Indeed, Richardson (1997, 43) claimed that sustainable development was 'a vague, contradictory, even meaningless concept,' 'a political fudge' and 'an expression of political correctness which seeks to bridge the unbridgeable divide between the anthropocentric and biocentric approaches to politics.' However, while its lack of clarity and contradictory potential has led some to claim that as a concept it is almost useless, Baker *et al.* (1997, 5) argued that 'Denying the usefulness of "sustainable development" as an analytic concept or the attractiveness of it as a normative concept . . . does nothing to impinge on either its popularity or importance as a political concept.' For Murphy and Bendell (1997, 35), sustainable development should be viewed as 'an emerging, positive myth, which has the potential to bring together diverse and often competing causes,' which could be 'an important catalyst for the formulation and implementation of creative and effective responses to many of the overwhelming challenges' facing the world today. This led Baker *et al.* (1997, 6) to suggest that instead of seeking a static and exact definition of sustainable development, we must instead understand it as a 'social and political construct' whereby its promotion 'forms part of a conscious process of achieving social change.'

The Place of Participation in Sustainable Development

For our purposes, one of the seminal features of the concept of sustainable development is the role that stakeholder participation plays in it. The achievement of sustainable development depends on political will (Pinto 1995, 73–73). But this political will is not confined to the level of the national state (Baker *et al.* 1997, 18–19). It is not only through top-down, centralized government action that sustainable development can be realized: as Matsui (1995, 70) explained, 'The right to sustainable development, as the successor to the right to development, implies the right of peoples to self-determination. Thus, participation of people and individuals in the process of sustainable development is an indispensable condition for its realization.' After the 1992 United Nations Conference on Environment and Development (UNCED), or Rio Earth Summit, the theme of participation and participatory policy-making was strongly advocated as a core prerequisite of sustainable development (Bruyninckx 2006, 268). *Agenda 21* (an outcome document of UNCED), in particular, 'places a strong emphasis on people and on their communities and organizations (including NGOs) in an approach which can be broadly described as "bottom-up" and which stresses the needs of the poorest' (Reid 1995, 187). Baker *et al.* (1997, 25) argued that the Rio Summit 'has given participation a new status,' and that 'Participation is no longer an optional extra.'

There were many reasons for this emphasis on participation, not least that increased stakeholder participation improved decision-making and policy implementation, because it integrated diverse perspectives to create a more holistic view of the relevant issues (Hematti 2002, 10). Furthermore, development professionals argued that past experience proved that for development to be sustainable it must be designed and implemented with the active involvement of the intended beneficiaries (Chauhan 1983, 8–9; Holmberg & Sandbrook quoted in Reid 1995, 233; Wise 2001, 56). Empowering people to make decisions and direct the course of their own development, was critical to successful implementation. Moreover, 'In contrast to a top-down policy approach, bottom-up participation has the potential to facilitate and catalyse radical social change' (Baker *et al.* 1997, 15). This was no less true of sustainable development policy in rich, than in poor, countries (Bruyninckx 2006, 267; Hemmati 2002, 25). In addition, 'the idea of participation as *empowerment* is that the practical experience of being involved in considering options, making decisions, and taking collective action to fight injustice is in itself transformative' (White 2000, 146).

It has also been claimed that participation endowed decision-making with greater legitimacy, reinforcing the principle of democracy. Bruyninckx (2006, 267) reported that 'A participatory society is believed to be a better society, as it fundamentally recognizes the role of citizens and social groups for the legitimacy of policy-making processes.' This argument was based on the premise that stakeholder participation 'is part of a significant development in democracy aimed at replacing one power with many' (Hemmati 2002, 45). Baker *et al.* (1997, 24) pointed out that this was not about building support for, and legitimising, pre-determined sustainable development policies conceived by top-down government initiatives, but about bottom-up community initiatives: 'Increasingly, participation is seen as a necessary part of the formulation and implementation of policy. . . Here participation becomes a process of respecting and drawing upon an indigenous community's own understanding of, and interactions with, the natural environment.'

The UNCED process itself was indicative of this new commitment to involve groups, beyond government, in sustainable development policy processes, because UNCED 'began a new era of inclusion' of NGOs at the international level, in processes which were traditionally the sole preserve of governments (Gallagher 2001, 341). As Reid (1995, 191) noted:

> Though barred from negotiations at these meetings they [the NGOs] contributed in a number of ways, which were generally welcomed by government delegations . . . Some NGO representatives were official members of government delegations, some negotiated on behalf of their governments, some had close links with the secretariat and some played a role in producing national sustainable development strategies.

Gallagher (2001, 341) reported that this participative 'trend has carried over into the many negotiations for international treaties and conferences that have occurred since then.' For instance, the Commission on Sustainable Development (CSD), established under chapter 38 of *Agenda 21* 'to ensure effective follow-up of United Nations Conference on Environment and Development' and to be 'responsible for reviewing progress in the implementation of Agenda 21 and the Rio Declaration' (UNDSD 2006a),

> has followed the example of Rio where NGO representatives outnumbered official government ones by two to one, and encouraged a considerable amount of NGO participation . . . The CSD's policy is an important consequence of Agenda 21's emphasis on the participation of a range of social groups and has meant that representatives of community organizations can, with the help from major NGOs, challenge other interest groups and influence decision-making at CSD meetings (Reid 1995, 203).

In fact, Hemmati (2002, 4) argued that at the international level, 'the most advanced multi-stakeholder discussions have been taking place at the UN . . . CSD . . . where there are well-prepared multi-stakeholder dialogues each year,' and 'Although the approach at the CSD is still evolving, it has become a model of multi-stakeholder engagement within the UN system on sustainable development issues.' However, while the participation of non-governmental stakeholders was embraced in the CSD, governments still retained their primacy as the central players in decision-making, while other actors merely fed into these processes. As Hemmati (2002, 61), citing Edwards, pointed out, almost all multi-stakeholder processes of this nature were designed 'effectively to give . . . voices not votes.' Dodds (2002, 34) further asserted that 'Stakeholders know they are not elected and are not asking for a seat at the table to vote on agreements. What they want is the opportunity to present their ideas and expertise.' Moreover, although the CSD was pioneering in this respect, the CSD's approach to stakeholder involvement 'is ad hoc and is at the discretion of the chair of the CSD' (Dodds 2002, 34).

Nevertheless, *Agenda 21* affirmed that while sustainable development was 'established as an overarching concept in intergovernmental processes, [it] is not something that governments and agencies can achieve on their own. It takes the contribution of all stakeholders and their individual and collective actions to bring about the changes required'

(Hemmati & Whitfield, 2003a, 3). Hence, *Agenda 21* identified and discussed the role of nine "major groups" in the implementation of sustainable development: workers and trade unions, farmers, women, youth, non-governmental organizations, the science and technology industry, local authorities, business and industry, and Indigenous Peoples (Hemmati 2002, 3): 'It recognized the need to engage these "stakeholders" in the development, implementation and monitoring of the global agreements' (Dodds 2002, 28).

Although some writers have criticized this stakeholder categorization for its exclusion of other groups, such as the academic community and the media (Dodds 2002, 37), Hematti (2002, 3) pointed out that, no matter what the shortcomings, 'Agenda 21 is the first . . . UN . . . document to address extensively the role of different stakeholders in the implementation of a global agreement.' Moreover, Rio appealed for the creation of a global partnership 'based on new levels of co-operation between all key sectors of society and government,' a principle which was broadly accepted internationally (Murphy & Coleman 2000, 209). Indeed, Murphy and Bendell (1997, 10) asserted that, 'partnership has become the cornerstone of post-Rio implementation of sustainable development.' Since Rio, at least at the international level, participation by non-governmental actors has been encouraged, and the United Nations has continued to work towards involving "major groups" in sustainable development, especially in the work of the Commission on Sustainable Development.

On the other hand, just as there were doubts raised over the commitment of governments to participatory processes after the Brundtland Report was produced, so, too, were questions posed after Rio and *Agenda 21*. Indeed Reid (1995, 197) asked: 'How seriously are governments committed to "bottom-up" approaches, participation by NGOs and "open governance" with its emphasis on participatory democracy and accountability, all of which entail some sharing of power if seriously implemented?'

Ten years after Rio, the 2002 World Summit on Sustainable Development (WSSD) was held in Johannesburg, South Africa and 'more than 22,000 people participated . . . including more than 10,000 delegates, 8,000 NGOs and representatives of civil society, and 4,000 members of the press' (UNDESA 2002a). This was 'the first major environment and development conference to have formally structured official input from a wide range of "major groups" of stakeholders . . . rather than relying upon the unofficial "side events" to provide a proxy input from global civil society' (Seyfang & Jordan 2002, 21). Indeed, the United Nations Department of Economic and Social Affairs (UNDESA 2002a) maintained that 'there was a new level of dialogue in Johannesburg between all the stakeholders, especially between governments, civil society and the private sector. Beyond speeches and platitudes, the participants in the Summit were forced to confront the needs and the arguments of other actors in a truly interactive dialogue.' Civil society groups' 'views and experiences were given prominence' in new ways at WSSD, with major group representatives being asked 'to sit on expert panels as well as make statements from the floor,' and to take part in 'plenary discussions, thematic dialogues' and 'high-level roundtables' (UNDESA 2002c). Furthermore, the UNDESA (2002c) contended that during interactive sessions at the Summit, major groups were recognized 'not only as stakeholders but also as accepted and respected partners in sustainable development.' Thus, the WSSD was seen by some as a seminal juncture in the establishment of participatory approaches for sustainable development at the international level, leading the UNDESA (2002c) to conclude that, 'The meaningful inclusion of civil society in this major summit represented a significant step forward for public participation in the sustainable development process.'

The Emergence of Multi-Stakeholder Partnerships for Sustainable Development

At the WSSD, the concept of sustainable development was still central, though increasingly contentious (Bruyninckx 2006, 270). The Summit produced two important documents, the short political *Johannesburg Declaration on Sustainable Development* (JDSD), and the longer action-orientated *Johannesburg Plan of Implementation* (JPOI), which included 'new global sustainability targets and some more generalized commitments' (Ayre & Callway 2005, 15). However, 'these classical products of multilateral consensus diplomacy' were not the only outcomes of the Summit (Pallemaerts 2003, 283). Alongside the traditional, formal governmentally-agreed commitments (Type One outcomes) was the advent of voluntary, multi-stakeholder partnerships between governments and other stakeholders from civil society and the private sector

(Type Two outcomes, or TTPs), aimed at complementing governmental efforts to implement the commitments made in the JPOI, *Agenda 21* and the *Programme for the Further Implementation of Agenda 21* (UNDSD 2003). Pallemaerts (2003, 283) pointed out that what was new about this development was less the fact of cooperation between states and non-state actors than 'the special prominence given to them in the context of an intergovernmental political process.' Furthermore, 'While partnership mechanisms pre-date the World Summit on Sustainable Development (WSSD) . . . the WSSD was the first time when there was official recognition that such partnerships could complement the work of governments in meeting the goals outlined in the international sustainable development agreements' (UNCSD Secretariat 2005).

Thus, while these TTPs were not a "formal" outcome of the Summit, they were endorsed as key mechanisms for the implementation of the agreements made by governments at Johannesburg. As Ayre & Callway (2005, 15) contended, 'If the Rio Earth Summit gave us the idea of involving stakeholders to deliver better informed decisions, then Johannesburg recognized that stakeholders should be involved in the practical delivery of the global agreements.' Indeed, the JPOI explicitly stated, in paragraph three of its introduction, that 'the implementation [of the outcomes of the WSSD] should involve all relevant actors through partnership, especially between Governments of the North and South, on the one hand, and between Governments and major groups, on the other' (UNDESA 2002b).

Partnerships were encouraged to register with the CSD Secretariat, because the JPOI expressly charged the CSD with the task of serving 'as a focal point for discussion of partnerships that promote sustainable development, including lessons learnt, progress made and best practices' (UNDESA 2002b). The United Nations Department of Economic and Social Affairs reported that 'Over 220 partnerships (with $235 million in resources) were identified in advance of the Summit and around 60 partnerships were announced during the Summit by various countries' (Pallemaerts 2003, 283). In the years since the WSSD, the number of new TTPs being initiated continued to grow, and in January 2008, there were 334 partnerships officially registered with the CSD (UNDSD 2008a).

The extent of enthusiasm for the new TTP approach as it emerged as an instrument of both global governance and governance at lower levels, was illustrated in an observation made by Jonathan Lash (quoted in

UNDESA 2002a), the President of the World Resources Institute, who declared of the WSSD that:

> This Summit will be remembered not for the treaties, the commitments, or the declarations it produced, but for the first stirrings of a new way of governing the global commons – the beginnings of a shift from the stiff formal waltz of traditional diplomacy to the jazzier dance of improvisational solution-oriented partnerships that may include non-government organizations, willing governments and other stakeholders.

It was claimed that the TTP framework would help overcome past inequalities between the North and South within development projects, moving away from donor-driven prescriptions, and opening up a space whereby representatives from developed and developing countries could devise strategies together. As Nitin Desai, the Secretary General of the WSSD (quoted in UNDESA 2002a) said, 'For those of you who have worked in developing countries, you are always at the receiving end of prescriptions and conditionalities. We need a shared programmatic structure framework and the partnerships help meet this need.'

However, critics saw three possible dangers in the TTP mode of implementation. First, it could weaken intergovernmental aid efforts. The fact that the partnership approach was strongly promoted by the UN and certain government delegations, such as the USA, in preparatory meetings for the WSSD, led some commentators to suggest that in doing so, these actors were 'paradoxically, downplaying the significance of intergovernmentally negotiated results and lowering the level of public expectations with respect to such traditional "first type outcomes" of United Nations conferences' (Pallemaerts 2003, 283). Although the *Guiding Principles for Partnerships for Sustainable Development* (Kara & Quarless 2002) (which were drawn up in Bali at a preparatory meeting for the Summit) specified that 'Partnerships are to complement the intergovernmentally agreed outcomes of WSSD: they are not intended to substitute commitments made by governments,' there was anxiety 'about governments off-loading their responsibility to the promise of partnerships' (Küpçü 2005, 95).

Second, as TTPs developed out of the work of the CSD, there was a danger that they would reflect the division in the CSD model of participation between governments as the core players and central decision-makers, and other stakeholders who had an input into these processes, but who were not considered to be equals. In other words, TTPs could end up as partnerships between governments with stakeholder involvement, rather than as genuine partnerships between all stakeholders.

Third, there was a fear that TTPs would not be properly audited. The *Guiding Principles* for TTPs stipulated that partnerships 'should specify arrangements to monitor and review their performance against the objectives and targets they set and report at regular intervals ("self-reporting"). These reports should be made accessible to the public' and partnerships 'should [also] keep the Commission on Sustainable Development informed about their activities and progress in achieving their targets' (Kara & Quarless 2002). This requirement was reinforced in the year following the WSSD when the CSD met for the 11th time and agreed that partnership reporting should be 'transparent, participatory and credible' and completed regularly, 'preferably on a biennial basis' (UNDSD 2003). But a system of self-reporting was inadequate for auditing purposes. Although TTPs were promoted by the UN, registered with the CSD, and thus were essentially endorsed by the UN, there was no clarity on how their effectiveness, transparency, and accountability could be ensured (Pallemaerts 2003, 286). We will discuss these three possible dangers related to TTPs in greater detail in the next chapters.

Governance for Sustainable Development

Notions of the role of participation in achieving sustainable development are related to the wider notion of "governance". Tolentino (1995, 137) argued that a 'significant aspect of . . . [our] rapidly changing world is the great concern about popular participation for good governance in sustainable development.' While the term "governance" has been defined in many different ways, it is almost universally acknowledged to involve the widening of participation beyond politicians and civil servants to other stakeholders in sharing responsibility for managing public affairs (CGG 1995, 2; Huillet 2004, xvi; Lovan *et al.* 2004, 7–8; Pierre 2000a, 3–4; Weiss 2000, 800). Hence, 'Governance includes the state, but transcends it by taking in the private sector and civil society' (UNDP 1997, iv).

Governance has a global dimension. 'Global governance is a political response to economic, cultural, social and ecological globalization' (Biermann 2006, 248). The notion of global governance refers to the fact that world politics (as well as national politics) is no longer solely the domain of nation states (Rosenau 2000, 187–188). Stakeholder participation also operates at the global level and 'stakeholders play an increasing role in implementing what has been agreed at international level' (Dodds 2002, 27). This includes the private sector, and although nation states are to some extent constrained by 'the workings of global markets in which large corporations and private financial interests exercise great power' (de Alcántara 1998, 111), the value of engaging in governance processes for sustainable development has become increasingly apparent to the private sector, largely for two reasons. First, they hope to have some influence over policy decisions that will affect their businesses in terms of regulation practices and the enforcement of governmentally-agreed environmental commitments (Hemmati 2002, 25). Second, they can ensure they are seen to be acting in a socially responsible manner, since a company's responsiveness to social and environmental matters can seriously affect share prices and profits globally (McLaren 2002, 1). Furthermore, the JDSD (UN 2002, para. 27) clearly stated that, 'in pursuit of its legitimate activities the private sector, including both large and small companies, has a duty to contribute to the evolution of equitable and sustainable communities and societies.'

Stakeholder groups, including NGOs concerned with the environment and human rights, networks of experts, as well as international courts and intergovernmental organizations, also now have a role in world politics (Biermann 2006, 242). Indeed, the participation of civil society actors in governance at the global level is seen as a counter-balance to the highly influential roles that multinational corporations and private financial interests assume, as well as a means of encouraging transparency and accountability of states (Pratt 2004, 315). As the European Commission (2001, 15) argued, 'Participation is not about institutionalising protest. It is about more effective policy shaping.'

One characteristic of global governance has been the emergence of new institutions beyond formal binding agreements between states, involving networks, partnerships and cooperation between non-governmental and governmental organizations (Biermann 2006, 248; Jessop 1997, 575; Pierre 2000a, 3; Rhodes 2000, 54). This trend is also seen at national and local levels, where states have moved away from direct service delivery, and have instead engaged in cooperative endeavours, networks, and partnerships with NGOs and the private sector (Grindle 2004, 541; Jessop 1997, 575; Pierre 2000b, 243; Poncelet 2004, 26). Teisman & Klijn (2002, 198) contended that 'Partnerships are seen as the best way, in the end, to govern the complex relations and interactions in a modern network society.' Instead of being one centralized decision-making body, there is 'an amalgam of centres of authority at various levels' (Biermann 2006, 248). Essentially therefore, as Dodds (2002, 37) argued, 'We are witnessing the recognition that, in a highly complex, globalizing and interdependent world, governments no longer have the power and ability to forge and fully implement all the various agreements that they conclude.' No longer can the state alone provide the solutions to societal problems (Hall 2005, 112).

This changing role of government has been a response to worldwide societal problems such as 'environmental pollution, currency crisis, corruption, AIDS, terrorism, mass migrations, and the drug trade' (Rosenau 2000, 172–173), all of 'which require broader sets of approaches and instruments' than those made available by the traditional state-centred model (Kooiman 2000, 139). As Kooiman (2000, 142) asserted, 'No single actor, public or private, has the knowledge and information required to solve complex, dynamic, and diversified problems; no actor has an overview sufficient to make the needed instruments effective; no single actor has sufficient action potential to dominate unilaterally.' On this view, 'The UN's continued legitimacy will depend on its extension of rights to civil society and other external constituencies to complement its base of inter-state relations and rights . . . based on the recognition that the implementation of state decisions can only be achieved in partnership with a range of constituency groups' (Secretariat of the Cardoso Panel 2004).

However, this does not imply that the role of government is on the decline or shrinking: the functions of governments were merely changing and adapting (Kooiman 2000, 139; Pierre, 2000a, 3; Pierre 2000b, 242–3). As Haque (2004, 271) suggested, the recent 'emphasis on partnership between the state and non-state entities has emerged in line with the . . . shift in the government's role from rowing to steering.' Similarly, Poncelet (2004, 26) stated that, 'Public authorities are being asked to . . . serve as facilitators and brokers of economic and social activity rather than as commanders and controllers.' Both Wise (2001, 57) and Reid (1995, 234) emphasized the continuing importance of the role of the state, arguing that participation would only empower people and lead to effective results if it was supported by governments with actions, not just words, including guarantees of civil and political rights. As Edwards & Hulme (2000, 46) pointed out, 'The state remains the ultimate arbiter and determinant of the wider political changes on which development depends, and it controls the economic and political frameworks within which people and their organisations have to operate.'

Critiques of Governance for Sustainable Development

We have established that participatory mechanisms of governance, including multi-stakeholder partnerships, which embraced actors beyond the nation state, emerged as a prerequisite for the achievement of sustainable development. However, these participatory mechanisms were not without their critics. One criticism was about the promotion of unelected civil society participation in governance, and the consequent 'loss of public accountability of decision-makers' (Pratt 2004, 314; cf. Meadowcroft 2007). This raised the issue of whether there could be a system of global governance for sustainable development, and if so how viable such a system would be and what form it should take (Bruyninckx 2006, 276). The UN was the obvious candidate, because it was the core international arena for addressing sustainable development, and a potential 'institutional anchoring point' for a system of global governance for sustainable development (Bruyninckx 2006, 276). Seyfang and Jordan (2002, 22) asserted that 'In a fragmented world made up of over 200 sovereign states, the UN is probably the *only* effective forum in which the global dimension of common problems such as sustainable development can be adequately resolved.' Similarly, the JDSD (UN 2002, para 32) emphasized 'the leadership role of the United Nations as the most universal and representative organization in the world, which is best placed to promote sustainable development.'

Dodds (2002, 27) pointed out that while the UN was created on the basis 'of the supremacy of the nation state,' this "supremacy" was now contested, as the effects of globalization were being taken into account by the intergovernmental system. Indeed, the UN Secretary General, Kofi Annan (quoted in Dodds 2002, 27) declared in 1999, when addressing the World Economic Forum, that, 'The United Nations once dealt only with governments. But now we know that peace and prosperity cannot be achieved without partners involving governments, international organizations, the business community and civil society. In today's world, we depend on each other.' Dodds (2002, 29) also noted that, 'Through the 1990s, the reform packages that have had an impact on the UN and global governance have nearly all been accompanied by an increase in the role and responsibilities of stakeholders.' Furthermore, in his report to the Millennium Summit on the role of the UN in the twenty-first century, Kofi Annan (quoted in Mestrum 2003, 47) urged that the UN 'must be opened up further to the participation of the many actors whose contributions are essential to managing the path of globalization . . . civil society organizations, the private sector, parliamentarians, local authorities, scientific associations, educational institutions and many others.'

Thus the creation of the TTP framework indicated a step forward in the UN's institutionalization of stakeholder involvement 'in the design and implementation of elements of a global governance system for sustainability' (Bruyninckx 2006, 277). However, while some divisions of the UN were partners in some of the TTPs, there were many TTP initiatives which did not count any UN agency as a member organization. Moreover, while partnerships had to meet certain criteria in order to register with the CSD, thereafter they were left to self-report, and the CSD was not endowed with steering, monitoring or evaluating powers over TTPs. In 2004, the Panel of Eminent Persons on UN-Civil Society Relations convened a workshop on partnerships to inform its report to the UN Secretary General. One view that came out of this workshop was that partnerships were 'not something the UN can control' (Secretariat of the Cardoso Panel 2004).

The Concept of Partnership

The Aspirations of Partnership

The concept of partnership has been integrally related to the concept of sustainable development (Davies 2002, 190; Mol 2007, 217–218; van Huijstee *et al.* 2007, 76). Glasbergen (2007, 3) defined partnerships as 'collaborative arrangements in which actors from two or more spheres of society (state, market and civil society) are involved in a non-hierarchical process through which these actors strive for a sustainability role.' In this book, we focus on partnerships that embrace all three spheres of society, aspiring to achieve the goal of sustainable development through non-hierarchical processes. However, we stress that the link between partnership and sustainable development has to be demonstrated; it cannot be taken for granted (Martens 2007, 34). We take note of van Huijstee *et al.'s* (2007, 85) point that 'The concepts "partnership" and "sustainable development" are more clearly linked discursively than empirically . . . the widespread assumption that partnership contributes positively to sustainable development lacks evidence.'

The idea of "partnership" has gained significant political influence in recent years on a global level and within national and localized contexts (Murphy & Coleman 2000, 208). For example, when first elected in 1997, the UK's Labour government, led by Tony Blair, 'tied its colours firmly to the partnership mast, announcing its intention to move from a contract culture to a partnership culture' when implementing its policies (Balloch and Taylor 2001a, 3). McDonald (2005, 579) observed of the concept of partnership, that it offered a third way and more efficient alternative to the free market and the strong state: 'At the level of theory, partnership working has been presented as a critique of both market- and state-led forms of governance, while in policy discourse they are presented as offering the potential for a more resource-efficient, outcome-effective and inclusive-progressive form of policy delivery'. In development circles, bilateral donors and international financial institutions have also increasingly talked of "partnership" relationships, as distinct from those based on conditionality (IDSUH, IDSRU & CMI 2002; Kayizzi-Mugerwa 1998; Maxwell & Conway 2000). According to Manz (2007, 3), 'These partnership models appear as the new "mantra" shaping the discourse on global politics and are perceived by many as a new hope for multilateralism.' Martens (2007, 11; 20) referred to 'what we can now describe as a boom in global partnerships . . . Nearly all areas of the UN system have now undergone the partnership boom.' The OECD (2006, 3) observed that 'The use of the partnership approach is growing worldwide.'

The notion of public/private partnerships (PPPs) became a major part of New Labour's agenda in the UK, but, as Poncelet (2004, 4) pointed out, PPPs were essentially bilateral rather than multilateral partnerships:

'Traditionally public-private partnerships have been restricted to collaborations between either government and business or government and NGOs. Multistakeholder partnerships, by definition, involve actors from all three of these major sectors.' The rationale for PPPs was that 'the private sector is intrinsically more innovative and efficient than the public sector,' so by working through PPPs these skills could be harnessed without 'the profit motive gaining ascendancy over the public' (Stiles & Williams 2003, 403).

The rationale for multi-stakeholder partnerships includes innovativeness: indeed, the potential for them to spawn innovative solutions and new insights has been often asserted in the literature (Benner & Witte 2004, 45; Brinkerhoff & Brinkerhoff 2004, 255; Colin & Lockwood 2002; Dossal & Fanzo 2004, 333; Hematti & Whitfield 2003a, 3; Mackintosh cited in Balloch & Taylor 2001a; 1). But multi-stakeholder partnerships promised many other advantages. For example, according to Mitchell (2005, 124), by establishing a partnership with people affected by an initiative, 'socially acceptable solutions' were more likely to be identified, and 'a new sense of ownership of both problems and solutions, which leads to more effective and sustained implementation' could be generated. Moreover, by engaging with stakeholders, governments could also improve the legitimacy of their policies and their effectiveness, ensuring they were 'responsive to community needs' (Brewer & Hayllar 2005, 477). Although it could be argued that these benefits would not necessarily depend on partnership, and that other forms of engagement such as consultation would be sufficient, it was more likely that they would be realized if a partnership were formed (Balloch & Taylor 2001a, 1; McQuaid 2000, 21; Witte *et al.* 2003, 82).

It was claimed that individual partners often only saw part of a problem, whereas together in partnership a more holistic way of thinking became possible (Lasker *et al.* 2001, 184). Moreover, since each actor had different strengths and weaknesses, partnership could help partners achieve their own objectives more effectively, by disseminating best practice (Haque 2004, 272). In fact, Tennyson (2003, 31) suggested that institutional reform of partner's organizations might be a more important product of partnership than any other outcome. This related to Mitchell's (2005, 130) argument that 'an often assumed benefit of participatory approaches is that individuals and organizations learn.' In this sense, partnerships were transformative (Lasker *et al.* 2001, 185).

One of the major advantages claimed for multi-stakeholder partnerships was their potential for empowering the people they sought to help who were traditionally excluded from developmental processes (Haque 2004, 273; Lowndes & Skelcher 1998, 316; McEwan 2003, 472; Poncelet 2001, 277). White (2000, 153) reported that the participatory approach 'is founded on the assumption that those who have been excluded should be "brought in" to the developmental process. It represents the people in the bad, non-participatory past as passive objects of programmes and projects that were designed and implemented from outside.' Rowe & Devanney (2003, 380) claimed that partnerships which engaged local communities could 'bring both the benefits of local knowledge and experience and . . . develop social networks and capacity.' For example, in developing countries there were sustained attempts since the 1990s to use partnership between donor agencies and indigenous NGOs as a means of strengthening civil society so that it could act as a check on indigenous government and challenge the 'prevailing top-down institutional culture' (Lewis 2000, 254). Mohan (2002, 130) noted that agencies such as the World Bank had begun to place 'greater emphasis on "partnership" at all levels in an attempt to counter the tendency to dictate policy terms to recipient countries.' As Ingram (2004, XI) put it, it was 'a move from "You do as I say" to "We do what we agreed".'

The initiation of TTPs came about with much talk about the above benefits that a multi-stakeholder partnership approach could bring to international development and the achievement of the MDGs, beyond the benefits brought by traditional bilateral and multinational methods. One of the most obvious arguments for the promotion of TTPs was their potential 'to provide the multi-actor, integrated solutions required by the scope and nature of the problems being addressed' (Brinkerhoff & Brinkerhoff 2004, 254). Sustainable development issues such as poverty and protection of natural resources were fundamentally and intrinsically multi-stakeholder concerns because they could not be addressed by any single actor, and they had implications and consequences which affected so many different facets of society (Dossal & Fanzo 2004, 333; Hemmati & Whitfield 2003a, 3; Huxham 1996, 4; Stiles and Williams 2003, 403). In addition, TTPs were trumpeted as a modality through which new resources could be mobilized, especially by harnessing private sector funds and expertise, and utilising civil society's knowledge and skills (Balloch & Taylor 2001a, 1; Küpçü 2005, 95): 'drawing upon a multitude of experience, knowledge, understanding and insight' (Mitchell 2005, 124).

However, as Ivanova (2003, 24) pointed out, partnerships such as the TTPs were not designed to replace the role of government, which they could not. While the private sector and NGOs could turn their backs on the implementation of developmental projects, government could not relinquish its responsibilities (Haque 2004, 287). Lowndes & Skelcher (1998, 316) suggested that partnerships could be considered as 'complementing formal democratic processes.' Ivanova (2003, 24–5) argued that 'the expertise they mobilize complements governmental expertise, their speed of action and reaction induces faster governmental response, and their horizontal, cross-border sources of legitimacy complements the traditional vertical representation process and legitimacy of nation-states' (Ivanova 2003, 24–5). Schipulle (2003, 51) claimed that this complementarity came about 'where stakeholder coalitions exist that are willing to advance towards sustainable global solutions that go beyond the minimum common denominator governments are willing to agree upon.' This might not be the only way that complementarity could exist between partnerships and governments, but it indicated faith in the ability of partnerships to achieve more than governments alone could achieve.

The Principles of Partnership

From these aspirations of partnership, we can distil its essential principles. It is important that the principles of partnership are identified, because in order to assess the potential of TTPs as a tool in governance, we must know what it means to "partner". Brinkerhoff (2002a, 325) asserted that what distinguished partnership from other forms of collaboration 'is the notion of *mutuality*' which was characterized by 'horizontal (as opposed to hierarchical) coordination and accountability and as equality in decision making, as opposed to domination by one or more partners' (cf. Brinkerhoff 2002b, 15–17). As mutual enterprises, partnerships were established only to realize objectives which could not be met by a single actor working alone (Huxham & Vangen 2000, 293; Mitchell 2005, 126; Stigson 2004; Wilson & Charlton 1997, 10). This has been termed 'collaborative advantage' by Huxham (1996). The collaborative advantage created through partnership depended on a degree of synergy, to ensure that the sum was greater than its parts (Balloch & Taylor 2001a, 2; Byrne 2001, 244; Evans *et al.* 2004; McQuaid 2000, 11). As Mackintosh (1992, 211) put it, 'the scope for partnership is defined by the existence of potential synergy arising from differences between partners.'

Mutuality thus lies at the heart of partnership, and there are seven principles of mutuality that characterize partnership. These seven principles are: mutual benefit; mutual responsibility; mutual respect; mutual trust; mutual commitment; mutual objective; and mutual participation. First, mutual benefit: all partners expect to gain from involvement in a partnership; in partnerships, 'actors are bound together by the mutually supportive pursuit of individual and collective benefit' (Cropper 1996, 82). So partnerships are not selfless entities, as Benner and Witte (2004, 46) pointed out: 'Partnerships are about enlightened self-interest, not about charity.' Mitchell (2005, 127) added that 'If there is not the prospect of benefits for all partners, and if the benefits will not be distributed or shared equitably, the prospects for a sustained partnership are low.' However, along with the shared benefits of partnership also comes responsibility, and partners have to share in any risks associated with their joint venture (Allison 2002, 1540; Poncelet 2001, 276). As Runciman (2003, 13) put it 'Partners sink or swim together.'

In terms of mutual respect, partnership requires partners to value one another in their partnership activities (CIARIS 2004, 8; Hemmati 2002, 55). Mitchell (2005, 127) stated that 'Even when differential power is held by partners, all partners must be able to be involved, and feel valued.' In other words, all partners have to recognize and respect the different skills, resources, outlooks, and expertise that fellow partners have brought to the collaborative enterprise, because 'It is difficult to imagine how a partnership can achieve synergy unless its partners appreciate the value of the others' contributions and perspectives' (Lasker *et al.* 2001, 192). Furthermore, respect entails trust, which is another characteristic of an ideal partnership; without trust between parties there is no basis for a collaborative way of working. Trust, in turn, depends on transparency. Mitchell (2005, 128), quoting Viessman, contended that 'A necessary condition for establishing mutual trust is that partnering arrangements be open, frank and honest.' Moreover, McQuaid (2000, 21) claimed that partnerships help to break down stereotypical views that partners hold about other partners, especially from different sectors, 'making joint working easier and more efficient.'

With regard to mutual commitment, Bovaird (2004, 200) suggested that partnership must also be 'based on a mutual commitment (over and above that implied in any contract),' in order that individual and collective objectives can be realized. Similarly, Robinson (1999, 3) stated that partnership 'implies an ongoing commitment to being involved with each other in a working (and often changing) relationship.' If there is no sustained commitment to the partnership from participants, then relationships will become strained and the partnership could cease to exist. Commentators have also pointed to the need for a mutual objective to bind the actors together in their shared enterprise (Lewis 2000, 254; Poncelet 2001, 276; Torjman 1999, 12). Although different actors usually have different motivations for becoming involved in any given partnership, reflecting their individual objectives, partners have to agree on some mutual objective or aim for the partnership as a collective whole, in order that the partnership has an identity which shows a communicable purpose to the outside world (Austin 2007, 63). The final characterising principle of partnership is that of mutual participation: the partnership concept implies more than dialogue, co-ordination, alliance, coalition, co-operation, or consultation; it affirms the need for mutual decision-making (Robinson 1999, 3). As Balloch & Taylor (2001a, 2) insisted, 'Partnership reflects ideals of participatory democracy and equality between partners.' However, Brinkerhoff (2002a, 325) cautioned that 'a careful balance between synergy and respective autonomy' must also be achieved, because partnership identity should not subsume the individual identities of partners and hence weaken the comparative advantage made possible through bringing the skills and resources of different parties together. This caution leads us to consider the pitfalls of partnership.

The Pitfalls of Partnership

As Brinkerhoff (2002b, x) pointed out, 'The term *partnership* connotes positive feelings and values.' Similarly, according to Huxham (1996, 7), partnership as a concept, along with the associated terms of collaboration, co-operation, co-ordination, coalition, network, bridge and alliance, necessarily carried positive connotations, by contrast to terms like conflict, competition, co-option and collusion, which have more negative connotations (cf. Haque 2004, 273). But the terminology of partnership, and associated words, does not always point to such positive implications. For example, 'partners in crime', 'network of terror' and 'unholy alliance' all have highly negative overtones. Moreover, on a practical level, partnership incurs considerable costs, requiring 'additional start-up investments in terms of time, energy, and, by extension, money' (Brinkerhoff & Brinkerhoff 2004, 263). The process of partnering can be frustrating because the logistics of partnering as well as the decision-making processes take longer than if any one partner acts alone (Huxham 1996; McQuaid 2000, 22).

Furthermore, partnerships often fail: 'not all partnerships work' (OECD 2006, 3). Earlier, we noted that one principle of partnership is a mutual objective or collective goal. Unsurprisingly, therefore, 'A lack of clear aims is often cited as a major cause of the failure of partnerships' (McQuaid 2000, 22). Caplan (2003, 32) suggested another cause of failure – that the mission statements of partnerships can be so diluted and weak, because they have been agreed upon to accommodate everyone, that they mean very little. A further cause is that different partners can have differing understandings of what the mutual goals means (McQuaid 2000, 22), or different intentions in becoming partners (Stiles & Williams 2003, 400). Tensions can arise over not just different organizational objectives but different philosophies, such as a belief in the value of the market, different management practices, different procedures, and different sector-specific languages (Huxham 1996, 4; McQuaid 2000, 25). As Mariotti (1996, 19) argued, 'When cultures . . . do not match, it is much more likely that disputes will arise over the way interaction occurs, even if there is strong agreement about the desire for the partnership to succeed.' In addition, while partnership has strong connotations of consensus, often 'inter-agency working involves a high degree of competition among organizations' (Lowndes & Skelcher 1998, 326). Such deficiencies can leave some parties disappointed, and thereby weaken trust between partners (Balloch & Taylor 2001a, 7; Caplan 2003, 34; Ebers cited in Stiles & Williams 2003, 403; Huxham 1996, 5). And as Mariotti (1996, 69) noted, 'trust is hard-won and easily lost.'

The existence of unequal power relations between partnering organizations is another reason why partnerships founder, because they undermine the normative principles of mutual respect and mutual participation (Brinkerhoff & Brinkerhoff 2004, 264; Mayo & Taylor 2001, 39–40). Poncelet (2004, 174) drew attention to the danger of co-optation: that 'the nonconfrontational tendencies of multi-stakeholder partnerships makes them vulnerable to cooptation by actors – especially the more powerful political and economic players in society – who may want to use these collaborative processes to build support for or manage opposition to their own policies or projects' (cf. Brinkerhoff 2002b, 177–178). These observations have led some critics to suggest that partnerships have been employed as 'discursive constructs designed to nullify opposition to dominant interests via processes of incorporation and interpellation' (McDonald 2005, 579). Meadowcroft (2007, 197) claimed that 'to the extent that partnerships are participatory at all, they privilege elite, over democratic, forms of participation,' while Backstrand (2006, 300) held that 'partnerships reflect rather than transform relations of power in global environmental governance' (cf. Martens 2006). Partnerships could 'just as easily entrench and reproduce existing power relations' as they could alter them (White cited in Pearce 2000, 33). Andonova & Levy (2003, 19) asserted that 'Participation in the WSSD partnerships is uneven and mirrors rather than challenges prevailing norms. Disparities in power and priorities that have dominated intergovernmental discourse over the past decade are quite visible in these partnerships.'

This danger of co-optation comes especially from those partners whose financial resources bestow the greatest power, which can lead to 'organizational homogeneity with the loss of valuable, distinctive, organizational characteristics' (Carroll & Steane 2000, 51; cf. Mohan 2002, 141; and Salamon, cited in Bovaird 2004, 200). In other words, everyone follows the piper's tune. For example, in a partnership between a Sri Lankan NGDO (non-governmental development organization), and a donor agency, Brinkerhoff & Brinkerhoff (2004, 266) claimed that power and financial disparities led to a 'directive relationship, not a partnership, where the identity of the NGDO was transformed from social movement to public service contractor.' Hence the contention that in partnerships between organizations from wealthy countries and those in the developing world, the 'espoused egalitarian relationship implied by the language of partnering' frequently conceals the 'principal-agent reality' (Brinkerhoff & Brinkerhoff 2004, 264). Thus, 'donor budgets . . . continue to determine the terms of engagement' (Kayizzi-Mugerwa

1998, 223). Similarly, Barrow & Jennings (2001, 21) argued that, in order to gain funding from Northern development NGOs, NGO groups in the South were led to 'comply with an agenda set by their Northern partners and, in so doing, to internalise the specific discourse used by Northern NGOs to justify that funding.'

Elliot (quoted in Lister 2000, 229) asserted that, in north/south development partnerships, it is 'a dialogue of the unequal, and however many claims are made for transparency or mutuality, the reality is – and is seen to be – that the donor can do to the recipient what the recipient cannot do to the donor. There is an asymmetry of power that no amount of well-intentioned dialogue can remove.' Biermann *et al.* (2007a, 251) claimed that 'the leadership of partnerships lies predominantly with industrialised countries,' and that 'state actors and intergovernmental organizations dominate the partner population.' According to Backstrand (2006, 304), TTPs 'reassert' and 'reinforce' governmental authority.

On the other hand, as Huxham & Vangen (2000, 298) pointed out, those partners who seem to be weaker, could provide some vital resource other than finances, and their 'threat of exit' could be a powerful one. This leads us to another pitfall, stated by Brinkerhoff & Brinkerhoff (2004, 264), that developing country actors, be they from government, business or civil society, could use partnership membership solely to advance their own ambitions: 'donors may naively expect partnership behaviour, believing they have created the proverbial enabling environment. In fact, the reality may turn out to be that partnering is simply another avenue to access donor funding, albeit dressed up in the rhetoric of solidarity and mutuality' (Brinkerhoff & Brinkerhoff 2004, 265). Similarly Haque (2004, 286) noted that, because in Bangladesh, NGOs and the government were often in competition for donor money, partnerships between the government and NGOs were not founded on trust but on a mutual desire to obtain foreign funding. Ndegwa (1996, 22) observed that in Africa, because states often viewed NGOs as competitors for funding, this led to hostile relations, increased governmental efforts to control NGOs, and 'political jealousy', which undermined effective partnership. Haque (2004, 285) also suggested that beyond financial motivations, NGOs might use partnerships to get close to government in order to escape regulation and government controls. Indeed, Biermann *et al.* (2007a, 256) claimed that 'partners themselves tend to be the primary beneficiaries of their partnerships' (cf. OECD 2006,

24–25). Moreover, Andonova & Levy (2003, 24) reported that a review of WSSD partnerships revealed that 'only a small portion of the registered partnerships involved all significant stakeholders,' while Martens (2007, 40) claimed that 'important sectors of society . . . [were] systematically excluded from partnerships.'

These issues highlight a central dilemma for TTPs – how to strike a balance between the need for equality between partners, and the need for leadership. Huxham & Vangen (2000, 299) wondered 'how trust can be built and maintained in situations where, on the one hand, the power relationships are perceived to be uneven and, on the other hand, they are seen as too equal, because no individual or organization has any hierarchical authority to enforce action.' Tennyson (2003, 18) asked 'How does leadership find expression in a partnership paradigm without undermining the principle of shared responsibility?'

Along with shared responsibility, comes shared accountability in partnership, but partnerships are very difficult to audit: 'As each agency has sacrificed some of its sovereignty in joining the partnership, it can also claim that the partnership, rather than itself, is the accountable body – yet there is often no direct mechanism by which these partnerships can be held accountable in a proper fashion' (Bovaird 2004, 203). If proper accountability measures are not put in place, TTPs could become a 'law unto themselves, accountable to everyone through "stakeholder engagement", and so to no one in particular' (Zadek 2003, 10). This is a valid fear because, although during the WSSD, the CSD Secretariat was responsible for assessing the credibility of TTPs when they registered, it became the responsibility of partnerships themselves to voluntarily report on their activities to the CSD Secretariat and the wider public (Ayre & Callway 2005, 38). This led many, most notably the NGO community, to voice their unease that this was a very frail instrument for accountability and partnership monitoring and evaluation (Ayre & Callway 2005, 38; Witte & Steck 2003, 4). Some felt that it was up to the NGO community to police unsavoury activities, but this was 'a role which many do not feel they should be expected, or even have the capacity, to fulfil' (Ayre & Callway 2005, 38).

Another difficulty relates to the weakness of the linkages between partnerships and multilateral, bilateral, and national objectives. At WSSD, NGOs and some governments levied criticism at the lack of linkages between the Type One outcomes, such as the *Johannesburg*

Plan of Implementation, and the Type Two partnership initiatives (Ayre & Callway 2005): 'The result was that the two processes co-existed but without substantial interaction or an exploration of their complementarities,' which led to a lack of clarity about the role for partnerships in global governance (Witte *et al.* 2003, 61). Benner *et al.* (2003, 88) insisted that 'partnerships can only make a meaningful contribution to the promotion of sustainable development if they can be tied into the overall international action agenda.' Unless TTPs are connected in a meaningful way to international and national policies such as poverty reduction strategies, it is argued, they will not maximize their potential (Malloch Brown 2004, 220).

Moreover, Hale & Mauzerall (2004, 221) noted that TTPs might weaken pressure on states to honour their intergovernmental commitments, and Ivanova (2003, 24–5) suggested that governments might use the smokescreen of partnership to conceal their lack of action. For instance, with regard to financing the goals agreed at the WSSD, critics saw partnerships as a means of diverting attention away from the limited amount of money that wealthy countries were spending on international development, a concern that was voiced by a coalition of developing country governments at the WSSD (Witte *et al.* 2003, 60). The United States spends only 0.1 per cent of its GNP on donor assistance and Germany only 0.27 per cent, even though both have agreed to the international target of spending 0.7 per cent GNP on donor aid (Sachs 2004, 206–7). As Jeffrey Sachs (quoted in Benner & Witte 2004, 45–6) noted, 'If there isn't real financial help and new financial help from the rich countries, these problems are not going to be solved in the poorest of the poor countries, no matter what partnerships are signed.'

Others have suggested that the promotion of TTPs by governments at the WSSD – especially the USA (Bruch & Pendergrass 2003) – was an attempt to avoid signing any new binding commitments (Witte & Steck 2003, 3; Witte *et al.* 2003, 60). These criticisms reflect a grave concern that governments have used partnerships to abandon their responsibilities and disperse them to other actors (Ayre & Callway 2005; Backstrand 2006; Haque 2004, 285; Küpçü 2005, 94; Martens 2007, 47; Meadowcroft 207, 198, 298; Witte & Steck 2003, 3). This is not only a charge that rich governments would fail to live up to their commitments, but also that within developing countries, partnerships would be utilized by indigenous government to transfer or diffuse blame to other actors for what is essentially indigenous government failure (Haque 2004, 285). As Mohan (2002, 149) suggested, 'partnership becomes an insurance policy against lack of effectiveness.' Bruch & Pendergrass (2003) argued that voluntary partnerships are insufficient to advance sustainable development, and that instead, international law and enforcement institutions need to be strengthened. Partnerships are not enforceable; do not have secure funding; and are restricted in their geographical reach.

As far as the role of the private sector in multi-stakeholder partnerships, and more specifically TTPs, is concerned, several issues have been raised. One deficiency was the lack of funding provided by the private sector for TTPs (Küpçü 2005, 95). As we saw earlier, one of the perceived advantages of multi-stakeholder partnerships over bilateral arrangements was a belief in their ability to draw on private sector funding so that more money would be directed towards sustainable development over and above that which governments were providing. However, in 2006, UN figures suggested that the private sector was only committing funds to approximately 20 per cent of partnerships (UNESC 2006a, 13). Martens (2007, 43) found that of the resources pledged to TTPs up to 2004, only 0.9 per cent came from the private sector.

Another charge was that, although advocates of private sector engagement argued that TTPs could be an important modality through which businesses demonstrated their commitment to socially responsible ways of working, not just profit-making to please their shareholders (Witte *et al.* 2003, 82), the reality was that TTPs offered the private sector a mechanism through which they would *appear* to be working for the common good yet still operating 'unsustainably' through exploiting the environment and society in their daily profit-making enterprises (Benner *et al.* 2003, 88). In other words, as some commentators have termed it, they would employ 'bluewash' or 'greenwash' tactics (Hale & Mauzerall 2004, 223; Küpçü 2005, 95; Witte *et al.* 2003, 60).

It was also claimed that TTPs offered the private sector a means of gaining valuable inside knowledge and thereby helped create for them new business opportunities and/or influence over policy outcomes and agendas (Carroll & Steane 2000, 50–51). Stiles & Williams (2003, 400) argued that PPPs 'can enable companies to leap-frog regulatory boundaries, enter new markets, acquire new skill sets, develop complex new product/service offerings, and facilitate multiple options to be explored relatively easily.' Mark Malloch Brown, former Chef de Cabinet to the UN Secretary-General and former United Nations Development Programme Administrator (2004, 215) contended that 'Companies are . . . beginning to see enlightened self-interest in opening up new markets in the developing world and the knock-on-benefits of developing new business models that can serve poor populations at the "bottom of the economic pyramid".' The fear was that instead of benefiting the poor, TTPs would be usurped as a tool through which a neoliberal agenda was promoted and privatization was made more palatable to developing countries. Indeed, TPPs would become 'strategic instruments' in the process of corporate capitalism (Farazmand cited in Sarker 2005, 251). Manz (2007, 3) referred to 'the risk of an overwhelming influence of private business.'

As we learned earlier, TTPs were much trumpeted for their multi-stakeholder approach, one of their aspirations being the involvement of civil society, entailing empowerment of those whom they sought to help, by giving them a say in the decisions which affected them. However, because of their voluntary nature, TTPs are not compelled 'to engage or represent the constituencies they affect,' nor are they necessarily answerable to beneficiaries (Küpçü 2005, 95). Furthermore, partnerships between northern donors and recipient country NGOs in rural development, which do not fully involve central and/or local government officials, have been criticized for not empowering rural people, but merely transferring service delivery away from government to non-elected NGOs, who are not representative of the people and are thereby unaccountable (Haque 2004, 284). Sarker (2005, 264) claimed that, in some cases, 'NGO groups act as simply a delivery mechanism, not as structures that could empower the poor.' Clark (1995, 597) pointed out that in terms of representation, Southern 'NGOs may represent a very narrow constituency, such as one kinship group, or the poorest farmers,' and hence cannot be assumed to speak for society as or a whole, or even for a community or beneficiary group. Furthermore, Collier (2000) maintained that, 'By providing goods or services directly to the poor, NGOs can reduce the accountability of local government to these people, undermining the foundation upon which future and long-term improvements in their lives must be built.'

Multi-stakeholder partnership has been proffered as a means of overcoming the global democratic deficit which exists as a result of neoliberal globalization whereby global governance 'Reflects the role and interests of a narrow band of humanity – "the wealthy and well organised"' (Pasha & Blaney 1998, 432). But cynics have argued that it is not clear how partnerships between 'essentially unrepresentative organizations – international organisations, unaccountable NGOs and large transnational corporations' would make global governance systems any more legitimate or representative (Ottaway quoted in Benner & Witte 2004, 36). Indeed, Lowndes & Skelcher (1998, 316) argued that instead of being a panacea for the democratic deficit, partnerships could create a situation 'in which non-elected bodies and self-selected representatives gain power at the expense of elected politicians'. In other words, partnerships are 'suspect because they dilute political control over decision-making' (Bovaird 2004, 200).

Another critical perspective often espoused by those opposed to neoliberalism and capitalism, is that partnerships with an empowerment agenda, or at least a participative approach towards the poor, in practice often merely provide some means of 'influencing the implementation of strategies that have already been decided on' (Byrne 2001, 256). Balloch & Taylor (2001b, 284–5) suggested that 'partnership becomes a tool of the established system for incorporation,' because it fails to address this issue of power. For example, Mosse (2003, 58–59) discussed a rural agricultural development project in western India where 'participatory techniques allowed the development priorities conveyed by the project (or demanded by its systems) to be mirrored back to them; in fact it was unavoidable.' In other words, the villagers merely endorsed the donor priorities and assumptions:

> And as villagers shaped their needs and priorities to match the project's schemes and administrative realities – validating imposed schemes with local knowledge and requesting only what was most easily delivered – the wider institutional interests of the project (including those of host and donor organisations) became built into community perspectives and project decisions became perfectly 'participatory' (Mosse 2003, 59).

Glasbergen (2007, 19) claimed that the 'WSSD partnerships are largely supply-driven (by what powerful actors have to offer) rather than demand-driven (by what is needed to bring a more sustainable development forward)' – a claim echoed by Andonova & Levy (2003, 19; 23; 26), Backstrand (2006: 299), and Hale & Mauzerall (2004, 233).

Thus, the rhetoric of partnership could undermine the autonomy of the communities that they were ostensibly benefiting. Fowler (quoted in Crawford 2003, 142) suggested that "partnership" in international development could be used as 'an instrument for deeper, wider and more effective penetration into a country's development choices and path . . . By appearing to be benign, inclusive, open, all embracing and harmonious, partnership intrinsically precludes other interpretations of reality, options and choices without overtly doing so.' Many authors reported on the use of partnership discourse by northern donors as a means of making their donor strategies more acceptable to recipient countries – i.e. in appearance, less top-down than previously (Brinkerhoff & Brinkerhoff 2004; Craig & Porter 2003; Greenhill & Wekiya 2004; Kayizzi-Mugerwa 1998; Lister 2000; Mercer 2003; Mohan 2002, 147; Mosse 2003, 58–59). For these authors, partnership was viewed 'as being a Northern-imposed idea which is deeply tied-up with the need for Northern aid agencies and NGOs to establish a legitimacy for operations in the South and demonstrate their "added value" in the development process' (Lister 2000, 229). For such commentators, partnership rhetoric with its focus on 'participation and good governance' could 'obscure a more covert and insidious expression of power which simultaneously empowers and normalises the actions of development partners' (Mercer 2003, 759). Thus, as Mercer (2003, 743) asserted, partnerships could 'serve to conceal continued conditionalities,' rather than achieve their espoused aims of empowering African states and civil society so that they have ownership over their own development. While the normative principles of partnership implied mutuality, partnership rhetoric could be used as 'a terminological Trojan Horse' to hide powerful interests and agendas (Fowler quoted in Crawford 2003, 142). Indeed, governments might partner with other stakeholders only superficially to garner legitimacy and defend themselves against criticism, because they 'want to retain their primacy within the process' (Teisman & Klijn 2002, 204).

Finally, Brinkerhoff and Brinkerhoff (2004, 259) showed how the appellation of partnership could be devalued by being applied indiscriminately, citing the fact that in the 1990s, the United States Agency for International Development (USAID) began to term any actor with whom it worked as a "partner", regardless of the nature of the relationship. This is a criticism which can also be directed at many other development agencies, leading Haque (2004, 272) to suggest that 'partnership has become a most frequently used buzzword in development debates . . . and perhaps an "overused and abused" term.' Andonova & Levy (2003, 29) described it as 'opportunistic window dressing', while Klitgaard (2004, 44–45) referred to 'Partnership Alchemy', and the 'partnership craze' within the World Bank.

Typologies of Partnership

The preceding discussion demonstrates that partnership can take many forms, and that as a concept, it is contested and ambiguous (Haque 2004, 272). Several writers have constructed typologies of partnerships to make sense of the concept. For example, Mattesich and Monsey (cited in Balloch & Taylor 2001a, 6) distinguished 'between types of joining up where partners maintain their individual authority but cooperate on some issues (usually at the margins of their main "business"), and types of joining up where partners pool authority.' Cropper (1996, 82) also identified this spectrum, terming it one of 'extremes of independence and fusion.' Likewise, Maxwell and Riddell (cited in Crawford 2003, 143) referred to 'weak' and 'strong' models of donor partnership. In weak partnerships, the relationship between participants is restricted to dialogue, the exchange of ideas, and the sharing of information. In strong partnerships, the relationship firms up into 'jointly agreed country programmes and multi-annual financial agreements' (Crawford 2003, 142). Hailey (2000, 315) reported Hately's distinction between partnerships in development with weak and strong senses of mutuality, which Hately termed 'conventional' and 'reciprocal' partnerships, respectively: 'Conventional partnerships are commonly short term, bureaucratic, one way and unequal . . . whereas reciprocal partnerships attempt to change the traditional way of working by creating two-way, horizontal relationships based on solidarity and equality.' As we shall see, both EUWI and WAWI have showed a tendency towards weak partnership, whereas PAWS has demonstrated a stronger sense of partnership.

A similar distinction has been drawn between partnerships as hierarchical or horizontal structures. For instance, The Centre for Informatic Apprenticeship and Resources in Social Exclusion (CIARIS 2004, 4–5) claimed that partnerships could take either a 'radial approach' or a 'networking' approach. In the radial approach, one or more actors promote a project, provide the initial funding, and take preliminary responsibility for administration of the initiative. These agents gather other actors in, and as a collective, they develop and plan the project. However, there is a 'danger in this model . . . that the sponsoring partner(s) will become overly dominant' (CIARIS 2004, 4). By contrast, the networking approach functions horizontally, with each partner having a similar input and gaining benefits according to their contribution. However, there are three dangers in the networking approach:

> First . . . because of the great amount of tension created by each stakeholder in quest of his/her own interests, there are no common denominators and the net . . . end[s] up fragmenting. Second, since there is no clear centre, the net will become a mess of useless knots and the action will not produce results, becoming irrelevant. Third, there is a risk that the net will continue expanding to such an extent that it will lose all direction and common strategy (CIARIS 2004, 5).

As we shall see, all three of our case studies adopted the radial approach, and in each of them there has been a tendency for the sponsoring partners to become dominant.

Focusing on the purpose of partnerships, Martelia and Schank (cited in Hailey 2000, 316) suggested a three-part typology:

> either a 'contracted resource', 'a way of working', or for 'acquiring critical consciousness'. Thus, if a partnership is seen merely as resource based it is about contracted delivery or tangibles with emphasis on cost-effectiveness and implementation. If the purpose is to develop a 'way of working' then it is intended to build linkages, decentralize responsibility, promote sustainability and empower staff and beneficiaries through shared decision making. The third type of partnership exists to promote a critical consciousness of the development process . . . [with Southern partners taking] more responsibility for the design and progress of the development process in which they are involved . . . [and Northern partners adopting] more of a 'consulting' role.

As we shall see, WAWI is closest to the contracted resource model;

EUWI is closest to the way of working model; and PAWS is closest to the acquiring critical consensus model.

Finally, partnership typologies have also been constructed on the basis of their styles of operation. For instance, Lewis (2000, 260) made a distinction between partnerships with an 'active' character; and partnerships with a 'passive/dependent' character. In an active partnership, the partnership evolves through negotiation and discussion which could sometimes result in conflict, but where the partnership is a learning experience and there are shared risks. Furthermore, it is an adaptive process inasmuch as 'although roles and purposes are clear they may change according to need and circumstance.' By contrast, passive/dependent partnerships have a more rigid blueprint type character, designed to implement specific projects. Passive/dependent partnerships are based on fixed attitudes and assumptions about the roles different types of organization could play, and about what value different actors could bring to a partnership. They also tend to be formed on the basis of 'availability of resources rather than on common objectives and shared risks' (Lewis 2000, 262).

As we shall see, EUWI and PAWS are active partnerships, whereas WAWI is a more passive/dependent partnership.

Evaluating Partnership

The monitoring and evaluation of TTPs is an area of considerable contention, yet no clear mechanism has been established to undertake this task: 'While the use of the partnership approach is growing world wide, there has been relatively little work done on evaluation . . . evaluation is not an established part of partnership initiatives . . . few evaluations have been conducted. There is no established methodology for conducting such evaluations' (OECD 2006, 11). Instead, partnerships have been left to report about their activities on a voluntary basis (Ayre & Callway 2005, 38). Some critics have argued that this is unsatisfactory, and that 'One of the key tasks for international organizations in the context of partnerships should be to create effective monitoring and evaluation structures' (Benner *et al.* 2003, 88). Müller-Kraenner (2003, 57) held that TTPs 'should not be allowed to become purely private endeavours, since they are, by their very nature, political animals that play an important role in the emerging system of global environmental [and developmental] governance.' Witte *et al.* (2003, 79) asserted that, 'If there are any rules that should be mandated for the proper functioning of partnerships, it should be rules for monitoring and evaluation.'

However, the need for an external evaluation system has not been accepted by everyone. For example, Björn Stigson (2003, 45), President of the World Business Council for Sustainable Development, urged caution:

> We should not strive to monitor voluntary initiatives. A few years ago, the Commission on Sustainable Development (UNCSD) went through a process of reviewing voluntary initiatives, but this proved unsuccessful. We should draw lessons from the past and favour a learning-by-doing approach. The value of voluntary initiatives is precisely in bringing together actors who believe they can achieve results by themselves and do not need an external body to monitor their actions.

But external monitoring might provide the only instrument 'whereby outsiders can arrive at informed judgements on the legitimacy or efficiency of a given partnership' (Witte *et al.* 2003, 78). Accountability is, in fact, one of the biggest challenges TTPs face, because 'if they are to be perceived as legitimate by outside observers and the public at large' then it is of paramount importance that they are transparent (Benner & Witte 2004, 41). This is not just a question of making those organizations and individuals involved in TTPs accountable, but 'it is crucial that the accountability of policy-making processes is also put to the test' (Benner & Witte 2004, 37). In order to allow such worthwhile evaluation, 'Internal procedures and governance structures have to be open to public scrutiny [as well as to partners] . . . Information on the internal division of responsibilities, voting rules and procedures – and most of all on funding (sources and spending patterns) – is crucial in this context' (Benner & Witte 2004, 40). Steets (2004, 2) added that there had to be transparency around 'the selection of participants, clear terms of engagement, proper reporting procedures and external evaluations' (cf. UNCSD Secretariat 2005).

This is not to suggest that the monitoring of partnerships is an easy task. Indeed, at the 13th meeting of the CSD (CSD-13) in 2005, participants 'highlighted the difficulties in defining metrics for success . . . it was also recognized that some partnership goals and targets . . . would require a more qualitative approach . . . [Thus], a flexible approach to monitoring and measuring partnerships' was advocated (UNCSD Secretariat 2005). One reason why partnership evaluation is a complicated and thorny process is that so many different types of relationship can be termed "partnership" that it makes it virtually impossible to draw up a checklist for successful partnership that can cover all cases.

Nonetheless, there are two key criteria that could be used in TTP evaluation: partnership as a way of working, and partnership as an instrument to achieve certain objectives. These criteria are not alternative options or even independent of one another: on the contrary, partnership is about a process, or a style of operation, initiated to achieve objectives, both collective and individual, be they essentially tied to the partnership method itself, such as empowerment of those normally excluded in development processes, or to more practical goals such as providing the population in a designated area with safe drinking water. Evaluation of partnerships therefore, has to cover *both* processes and outcomes. As Selsky & Parker (2005, 861) noted, in the 'normative literature', partnerships are viewed as ends in themselves, embodying ethical values such as participation and empowerment. But such values are also a means to the achievement of primary goods and services (Brinkerhoff 2002b, 17–18; Stern 2004, 39). Haque (2004, 279–280) pointed out the danger of interpreting outcomes too narrowly in terms of the internal functioning of TTPs:

> In assessing the outcomes of partnership, there is a tendency in the current literature to use certain criteria such as the level of each partner's satisfaction, effectiveness of conflict resolution, equity and accountability among partners, transparency in partnership activities, compliance of partners with mutually agreed contracts or obligations, and so on. . . However, these assessment criteria seem to consider partnership as an end in itself rather than a means to achieve the respective goals of individual partners. In other words, the ultimate success of partnership should be judged in terms of whether it has been conducive to the realization of the objectives of various partners, which they would not be able to achieve by themselves.

Similarly, Torjman (1999, 16) asserted that 'Partnerships never should be seen as an end in themselves: they are simply a means to achieve a certain goal.' However, while this is true, there is a danger in focusing solely on outcomes. The ability of a partnership to realize the objectives of its members rests on such qualities as equity, accountability and other ideal principles of partnership, because without these, partnership would not be possible. Therefore the process of partnership is important in determining a partnership's ability to fulfil its aspirations and realize its objectives. As Torjman (1999, 16) argued, 'the apparent obsession with outcomes-based evaluation has detracted from the equal importance of process.' TTPs have been initiated because of the perceived potential of the partnership modality in helping to achieve international, governmentally agreed goals. Therefore, in their assessment we must recognize the need to examine the processes of partnership as well as its outcomes.

As previously noted, it would be difficult to devise a checklist exhaustive enough to assess the success of all partnerships. Nonetheless, we can consider how our twin criteria of process and outcomes could be appraised. First, with regard to the *process* criterion, TTPs should meet the following procedural requirements (which we have compiled from our extensive survey of critical opinion in the literature):

1) Guarantee legitimacy, through agreeing the design of the partnership 'in a democratic, transparent and equitable manner, including the identification of stakeholder groups and participants, the framing of agenda and work plan' (Hemmati 2002, 60).
2) Involve 'all partners [or at least a good range of partners] from the outset (rather than the traditional sub-contracting approach)' (UNCSD Secretariat 2005).
3) Be 'inclusive and not exclusive' (Hemmati 2002, 59). If logistical and functional constraints make selection criteria necessary then these criteria need to be made public and open to debate.
4) Ensure there is a broad forum of stakeholders, in some cases involving 'those not necessarily part of project implementation but interested in or impacted by the partnership project' (Caplan *et al.* 2001).
5) Clearly define roles for all stakeholders which allow for flexibility, but have clarity about 'who is responsible for what and when' (Evans *et al.* 2004, 14).

6) Realize a 'common understanding of the intent and outcome of the process' (IDSUH, IDSRU & CMI 2002, 8).
7) Be willing and have the capacity to learn, evolve and adapt to changing circumstances.
8) Hold regular meetings and build transparent and 'solid decision-making structures' which embrace horizontal as opposed to hierarchical coordination and accountability (Caplan *et al.* 2001).
9) Ensure equitable participation of all partners, or at least an environment conducive to equitable participation even if one or more "partners" choose not to engage as fully as others do.
10) Involve intended beneficiaries, avoiding top-down prescriptions.
11) Attain mutual understanding of individual partner organizations motivations and constraints (Evans *et al.* 2004, 14).
12) Achieve an appropriate balance between mutuality and autonomy.
13) Create and maintain good communication channels, both internally to build relationships between partners, and externally to demonstrate effectiveness, attract new partners and funding, and improve legitimacy (UNCSD Secretariat 2005).

Second, with regard to the *outcome* criterion, TTPs should meet the following requirements (which again, have been compiled from the literature):

1) Realize objectives that no single actor could achieve alone and 'not merely reflect existing arrangements' (UNDSD 2003).
2) Be linked to global and national goals and strategies in order that their contribution is meaningful.
3) Avoid duplication of activities and enhance coordination between various stakeholders.
4) Provide innovative solutions combining the resources, capacities and knowledge of all involved.
5) Empower recipients and 'where relevant, result in . . . capacity building in . . . developing countries,' enhancing the effectiveness and sustainability of project outcomes through local ownership (UNDSD 2003).
6) Mobilize new resources over and above those already being supplied by governments.
7) Complement but not replace governmental efforts.

8) Avoid undermining the accountability mechanisms between citizens and their governments.
9) Meet the partnership's own espoused objectives.

Conclusion

This chapter has provided the conceptual framework which underpins the analysis of our three case studies of partnerships for water and sanitation supply in chapters three to five. This framework has two elements: the concept of sustainable development and the concept of partnership. In the section on the concept of sustainable development, we showed how the concept had evolved over the last 20 to 30, during which participation has emerged as an inherent characteristic necessary for the realization of sustainable economic, environmental and developmental agendas. As we explained, it is now almost universally acknowledged that sustainable development requires the active involvement of a whole host of stakeholders in its design and implementation. The Rio Earth Summit emphasized the involvement of non-governmental stakeholders in policy-making processes, and the Johannesburg WSSD highlighted the necessity for stakeholder participation in the implementation of sustainable development policies. These developments point to the fact that, in a globalizing world, governments no longer have the wherewithal to solve all of the societal problems they face, and instead of traditional hierarchical, top-down government, we are increasingly seeing shifts towards "governance", which is more inclusive than government and embraces non-governmental actors from civil society and the private sector.

In the section on the concept of partnership, we demonstrated that multi-stakeholder partnerships had the potential to mobilize new resources (not only, but including, those of a financial nature); to empower people whom the actions of the partnerships impacted upon; to identify socially acceptable solutions; to create a sense of ownership over problems and their solutions, leading to more effective and sustainable implementation; to bring about organizational and individual learning; to avoid duplication of activities; to provide innovative solutions through more holistic ways of thinking about a problem, or problems; and, through synergy, to achieve more than any single actor or organization could alone. For this potential to be fully realized, a partnership should uphold seven ideal principles: mutual benefit; mutual responsibility; mutual respect; mutual trust; mutual commitment; a mutual objective; and mutual participation. Rarely, if ever, has this ideal-type partnership been achieved, because in practice, partnerships could lack a common unifying objective; could include partners with incompatible individual aims; could fail to clarify individual motivations and risks at the outset, thereby weakening trust; could be unsuccessful in overcoming cultural differences between partners; could fail to empower those impacted upon; could raise significant fears about whom they are accountable to; could weaken public accountability; could reinforce unequal power relationships, rendering mutual respect and mutual participation difficult; could be co-opted by one or more partner; or could be used as a smokescreen to hide other interests and strategies (for example, governments could use partnerships to conceal inaction and disperse their responsibilities to other actors; NGOs could use them merely as a means to gain funds; and the private sector could exploit them as a means to gain access to new markets and 'greenwash' their activities).

At the risk of over-simplification, we may divide these potential dangers into two scenarios: first, the governance model could be used by government to preserve and strengthen its supremacy through superficially encouraging greater participation and engagement. In this scenario, a business-as-usual, bilateral/multilateral, top-down, government-centred approach would dominate TTPs, and the notion of partnership would merely be a veneer whereby governments (both of the north and south) would gain legitimacy and stave off criticism. The other potential danger lies in the opposite direction, that TTPs could offer governments an opportunity to abdicate their responsibilities and off-load them to other actors. This scenario has two strands. First, in the northern context, TTPs could be used by northern governments to deflect the pressure on them to increase their Official Development Assistance (ODA) and to take more affirmative action to reduce poverty in the developing world. Second, in the southern context, partnerships within an individual country could be used as scapegoats, to take the blame for what is really indigenous governmental failure (Mohan 2002, 149).

This study takes the view that TTPs must steer a course between these two potential dangers. Properly interpreted, governance allows for *both* government leadership *and* genuine stakeholder engagement and participation, so that governments may still retain a certain level of sovereignty and take ultimate responsibility, while other actors or stakeholders increasingly share responsibilities in the governance of public affairs. However, this is a difficult balancing act, because if governments remain prominent, and, to a degree, dominant, then it is difficult to envisage how equitable partnerships between state actors and other stakeholders can develop. In the cases studied in the next three chapters, this balancing act will form the central focus of our attention.

3

THE EU WATER INITIATIVE (EUWI)

Introduction

The EU is the largest provider of water-related overseas development assistance in the world. The EU Water Initiative (EUWI) was launched at the WSSD by the 15 Member States of the European Union (although the initial impetus came from the European Commission and the Environment Commissioner, Margot Wallström). EUWI was conceived as not only a partnership in itself, but also 'a platform for strategic partnerships,' and it was expected to continue until at least 2015 when the MDGs were set to be reached (EU 2002, 4). In the launch document, *EU Water Initiative: Water for Life – Health, livelihoods, economic development, peace and security. Implementing the Programme of Action of the WSSD* (EU 2002, 4), which was presented at the WSSD, the main objectives of EUWI were stated as to:

- Reinforce political will and commitment to action . . .
- Make water governance effective and build institutional capacity . . .
- Improve co-ordination and co-operation . . .
- Increase the efficiency of existing EU aid flows . . .

Since then, the objectives of the initiative have changed, and have been stated in varying forms in successive documents. The most frequently cited aims appeared on both the EUWI website (EUWI 2006a) and in the document entitled *Africa – EU Strategic Partnership on Water Affairs and Sanitation. Outline Strategy and 2004–2005 Work Programme* (AMCOW-TAC & EU 2003, 5), which sought to 'operationalise the conceptual document presented at the launch of EUWI.' Here they were reported to be to:

- reinforce political will and commitment to action,
- promote improved water governance, capacity-building and awareness,
- improve the efficiency and effectiveness of water management through multi-stakeholder dialogue and coordination,
- strengthen coordination through promoting river basin approaches, and
- identify additional financial resources and mechanism to ensure sustainable financing.

Although these two sets of objectives are very similar, one difference between them is that the more recent objectives moved away from the original goal of increasing the efficiency of EU development aid, towards the need to find alternative sources of funding.

In 2004, a European Commission official and EUWI Secretariat member (Interviewee 6, 2004) identified the aspirations of the initiative thus: 'The key objective . . . is to contribute to achieving the Millennium Development Goals and targets related to water by reinforcing political commitment and by increasing co-ordination and efficiency.' For another Secretariat member (Interviewee 7, 2004), coordination was the key factor:

> From my point of view the main word describing the initiative is 'co-ordination' . . . so we need to co-ordinate and organize the participation of everybody more efficiently . . . If we do not share the same point of view on how to work together it will be difficult to deliver so we need to share not only financial methodologies but also working methods.

Similarly, a third Secretariat member (Interviewee 11, 2004) held that 'there is a very strong harmonization agenda within the Water Initiative . . . The initiative is about frameworks . . . it's about co-ordination of efforts, it's not about funding projects because there isn't any money associated directly with the Water Initiative.'

EUWI is a broad partnership with a large and cosmopolitan make-up. According to its *Code of Conduct* (EUWI 2005i) it:

> is a tripartite partnership involving governments, civil society and water providers whether public or private . . . **Government** is involved through commitment to improve policy, governance, institutional capacity, public knowledge, education and awareness and safeguard sustainable access to water and sanitation, mobilisation of financial resources; improved management of water resources, and to contribute towards poverty reduction. **Civil Society** promotes advocacy, capacity building, mentoring, and better community organisation, and provides access to specific skills in service delivery and mobilisation of resources at local level. Civil Society contributes to the strength of the EU Water Initiative through feedback and better integration of communities' needs and perspectives and promotion of empowerment actions. **Water Operators**, both public and private, are partners in the EU Water Initiative because they are a primary source of managerial, technical and practical knowledge on serviceability standards and promotion of public awareness, as well as for the effective planning, operation, and maintenance of urban water infrastructure.

In fact, the multitude of participating organizations in EUWI is so complex that it is difficult to fit all the actors neatly into these three categories. The EUWI's website (EUWI 2004d) and the launch document produced for Johannesburg (EU 2002) revealed that its members included, first, at the European and international level, on the governmental side, the Member States of the EU; interested third countries; the European Commission; European local authorities; and other European institutions. Additional members included four international regimes, and from the private sector, six of the major European water companies, as well as various technical experts. There were also partners from European and international financial and development institutions. On the civil society side, at least eight organizations participated, and international water organizations were also involved. Second, at the regional level, membership of EUWI included, on the governmental side, potentially 53 African governments (in the form of the African Ministers' Council on Water [AMCOW] and the New Partnership for Africa's Development [NEPAD]); Latin American Environment Ministries; governments of the Mediterranean; as well as the national governments in individual partner countries. Other regional partners (potentially) included, were local authorities; water operators; NGOs; and experts in the individual countries where project work was undertaken.

EUWI is not focused solely on sub-Saharan Africa – it includes three other regional "modules" as well as three thematic "modules" – but our attention is restricted to its sub-Saharan African activities. Initially, two Working Groups (WGs) in Africa were established; one focused on Integrated Water Resources Management (IWRM) and the other on Water Supply and Sanitation (WSS). However, on 1 January 2006 these two groups merged to form the Africa Working Group, with Germany as the lead EU Member State. We have divided our discussion of EUWI into three contexts (although they are inherently linked): the northern context in terms of the partnership process and workings within the EU; the EU-Africa context in terms of partnership at the regional level between African actors and their northern counterparts; and the Zambian context as a case study of EUWI's work within an African country. Finally, the chapter will conclude with a consideration of the outcomes and achievements of EUWI in Africa.

The Northern Context

A Strategy for Development of the EUWI was written to 'create a common understanding of the EUWI and to guide action in future years' (EUWI 2005e, 1). In the final draft of this document, dated 26 September 2005, it stated that 'The EUWI will not achieve its intended objectives, unless EU partners, MS [Member States] as well as CSOs [civil society organizations] and interested public and private sector operators, share a sense of ownership and commitment to the EUWI mission' (EUWI 2005e, 4). This was an issue that EUWI struggled with over the course of its history, as we shall see.

The organization of EUWI was a constant source of deliberation during its first two years of existence. Initially, the structure consisted of a closed Expert Group made up of the Member States, to make political decisions; a Multi-Stakeholder Forum (MSF) led by the European Commission, to be used 'as a means to discuss and agree the final programmes of action' (EU 2002, 10); Working Groups, led by either individual Member States or the Commission, for each of the focal components, which reported to the MSF; and a small Secretariat staffed by the European Commission with the role of supporting and co-ordinating the partnership. However, this structure did not prove satisfactory, largely because the roles and powers of the different bodies were not well defined at the outset, especially those of the MSF (Interviewee 8, 2004; Interviewee 10, 2004; Labre 2004). Indeed, an NGO member (Interviewee 3, 2003) observed at the end of 2003 that while she had originally believed that the MSF was to be a decision-making body, proposals agreed at the MSF were regularly 'contradicted by Member States meetings.' She added that 'there is still, after two years, no structure, no decision-making process, no governance, nothing.' Annette van Edig (2004), a water advisor for the German Federal Ministry of Economic Co-operation and Development, similarly observed that for the first two years of the initiative, 'the Multi-Stakeholder Forum was . . . a waste of time because everybody was talking and nobody was steering, and in a way there was no leadership.'

Since June 2003, however, there has been a major effort to reform and clarify the structure of EUWI because, as one Commission official and member of the Secretariat, (Interviewee 7, 2004) noted in March 2004, 'now after two years of the design phase of this water initiative . . . we need to write our procedures, our rules, our way of working . . . on paper, very clearly, so that all Member States and stakeholders share the same view on this initiative.' After a year of work, and after a series of proposals had been debated, finally, in July 2004, the *Organisational Framework and Modalities for the EUWI* was adopted, with an agreed structure founded on four main bodies (Walshe *et al* 2004, 5): the Multi-Stakeholder Forum (MSF), which met annually, was open to all stakeholders, and had an advisory function; the Steering Group (SG), driver of EUWI; the Working Groups, in charge of implementation at the regional level and cross-cutting issues and which had now to report to the SG in addition to the MSF; and the Commission/Secretariat of the Initiative, which according to one Secretariat member (Interviewee 4, 2004) would 'ensure proper co-ordination and facilitate the advancement of the initiative and also do some of the day-to-day work that will be required to help the Steering Group in its task.' Three representatives from water operators and three from the NGOs were elected from the MSF's constituents to be members of the SG. The SG also included all interested Member States, European third countries, the European Commission, the European Investment Bank (EIB), chairs and co-chairs of the Working Groups and other invitees. It was to meet at least three times a year and had the responsibility to agree and ensure delivery of 'the strategic orientations of the Initiative . . . and relevant annual work programme of the EUWI,' and to take 'practical decisions about the operation of the EUWI' (Walshe *et al* 2004, 12).

This reorganization increased stakeholder engagement in EUWI decision-making. A Secretariat member (Interviewee 11, 2004) said in November 2004, that while, initially, the problem with EUWI's organizational framework was that:

> there was never a clear line of accountability within the structure . . . if you . . . compare the [new modalities] . . . to a business structure, we've now got a board of directors – the Steering Group, we've got the business units which are the Working Groups, and we've got the shareholders which is the Multi-Stakeholder Forum, and it starts to make a bit more sense in terms of lines of responsibility, of who does what, and of what the accountable bodies are.

Nonetheless, within EUWI, there is a feeling among many members that northern governments have dominated the initiative, in that EU Member States and the European Commission have assumed a superior status in relation to other stakeholders. As one NGO representative (Interviewee 10, 2004) put it, 'It's a very unequal partnership if it's a partnership.' Jacques Labre (2004) of Suez concurred, stating that 'everybody is not on an equal footing so I do not accept the word partnership in this case.' A member of the Secretariat (Interviewee 6, 2004) affirmed that EUWI:

> was initiated by the European Commission, and leadership is with the European Commission and EU Member States, that is clear. If you look at the Working Groups, they are all led by EU Member States. The whole process is led by public institutions with strong stakeholder involvement. It is a multi-stakeholder process but it is clear that the leadership of it remains with those who are in the end providing the funding instruments for the operational side.

Northern governmental dominance is further evident in the fact that while the partnership had no real funding of its own, northern governments provided the Secretariat in the form of Commission staff made up of nationally seconded experts from Member States (Interviewee 13, 2005). Although it is true that an independent consultant (Interviewee 10, 2004) argued that the Secretariat was 'under-manned and under-resourced in terms of power;' that Labre (2004) of Suez agreed that the Secretariat was weak; and that Monique Le Génissel (2004) of the French Department of Natural Resources and the Environment, and former leader of the IWRM component in EUWI, stated that the Secretariat did not have 'sufficient resources, they need many more people because . . . [they have] an enormous amount of work to do,' nonetheless, there was a broad consensus within the partnership that, as a Commission official (Interviewee 4, 2004) put it, the Secretariat 'is both a facilitator and a leader.' The primacy of the Secretariat was confirmed by another Commission Official (Interviewee 6, 2004), who asserted that 'it has been the driving force because it was the body at the centre of the process. It is a Secretariat with a strong element of leadership.' Similarly, an ex-EUWI co-ordinator and former Commission official (Interviewee 5, 2004) insisted that 'The Commission has always had a leading and driving role in the whole thing – that is very clear.' Some EUWI participants have suggested that in order to make EUWI less government-dominated with more equality and equity within the partnership, 'the Secretariat shouldn't be a Commission Secretariat but an EU Water Initiative Secretariat,' formed with the involvement of non-governmental stakeholders (Interviewee 10, 2004). In fact, NGO representatives offered to provide staff, but the Commission and Member States were reluctant to open up the Secretariat to those outside the Commission.

Opinion is divided over whether it was the Commission, the Council, or the Member States which wielded most power. Labre (2004) of Suez said it was the Commission that dominated: 'we [civil society and the private sector] are conscious that we don't have any decision power, but it is the same each time you deal with the Commission.' However, a member of the Commission and the EUWI Secretariat (Interviewee 7, 2004) said that power lay with the Council, stating that even with the new organizational framework for EUWI, 'to our minds the Expert Group [of Member States] is the beginning of the institutional process which is within the Council [of Ministers] . . . [and] all political and financial decisions will be taken within the Council.' A Member

State representative (Interviewee 13, 2005) said that power was shared between the Commission and the Council:

> The problem is that the EU Water Initiative is not a decision-making body, it is simply a structure that gives advice to the Commission and the Commission can only implement things that it presents to the Council for approval . . . The EU Water Initiative cannot make any decisions, and the Commission can only make decisions within certain limits, everything they do which is new is subject to approval by the Council.

Another interviewee (10, 2004) said power lay with the Member States, instancing how, up until the organizational changes in July 2004, the MSF had been by-passed by the 'so-called Expert Group' of Member States, which held separate meetings, and agreed policies without even the consultation, let alone the participation, of other stakeholders. By making decisions without involving other parties the Expert Group reduced the MSF to a mere talking shop, a place 'where you meet, you chat, you exchange views . . . not a process that you can make decisions and agreements in.' Indeed, even once the new organizational framework was implemented, the Expert Group, although no longer formally recognized within the EUWI structural framework, still existed informally. A Member State representative (Interviewee 13, 2005) justified this anomaly, on grounds that 'from time to time it is better that you have exclusive meetings because there are things that cannot be tabled when you have the private sector or NGOs present.' Similarly, a Secretariat member (Interviewee 4, 2004) argued that 'one of the aspects [of the initiative] is to make sure that we deliver EU aid in a more effective way, and that requires better co-ordination between the Member States and that is something which is clearly government based.' Moreover, because Member States provide financial support for the partnership, one member of the EUWI Secretariat (Interviewee 7, 2004) asserted that the Member States 'lead because they have money in their pockets.' Danielle Morley (2004) of the Freshwater Action Network argued that Member States assumed a more prominent role because at the end of the day, 'they have to take all the political decisions and all the financing decisions.'

These considerations reflect a deep-seated tension within the foundation of EUWI, which one independent consultant (Interviewee 10,

2004) identified with the following question: 'is the EU Water Initiative a European Union structure, or is it a partnership which involves others outside?' He answered that if ultimate authority over decision-making rests with the Council of Ministers, then it is 'an EU process, not a partnership process.' Darren Saywell (2004) of the Water Supply and Sanitation Collaborative Council (WSSCC), stated in early 2004 that 'You get the impression that they're going through the motions of consultation because that's what they need to be seen to be doing.' He argued that 'Partnership is a rather grandiose word for what you have . . . There isn't equality within the membership of the partnership.' A representative of a private operator (Interviewee 8, 2004) commented that essentially the task of EUWI was 'to balance . . . business as usual with the need to take the stakeholders into account at some stage.' Two members of the Secretariat (Interviewee 4, 2004; Interviewee 7, 2004) described EUWI as a 'multi-stakeholder *process*' rather than a multi-stakeholder *partnership*. Thus, terming it a "partnership" from the outset risked raising false expectations about how it would operate. As a not-for-profit representative (Interviewee 2, 2003) noted, 'the EU Water Initiative . . . built up a whole series of expectations' which it then couldn't control.

It appears, therefore, that EUWI has a two-tier structure, with partners at the top (comprising governmental bodies such as the European Commission, the Council and Member States), and stakeholders at the bottom (comprising non-governmental bodies like the private sector and NGOs). One Member State representative (Interviewee 13, 2005), describing MSF meetings, said that: 'there was an inner circle and an outer circle, the inner circle would be the Member States and the Commission and . . . we couldn't make decisions but we could decide what to recommend, and then you had an outer circle of multi-stakeholders . . . and they could voice their opinions.' A Secretariat member (Interviewee 4, 2004) stated that 'I can't agree that the Commission or Member States are stakeholders . . . what we understand as stakeholders are NGOs and the private sector.' A representative from a private water firm (Interviewee 8, 2004) maintained that:

> stakeholders are not partners in this partnership . . . there is a partnership between the African governments and the EU basically, so this is the partnership. And then we have the stakeholder

> process which is supposed to sort of surround this whole thing but we are not considered as partners . . . The Commission, the Member States, and the African governments are partners. The whole governmental level is where the partners are. Stakeholders are at the non-governmental level.

An independent consultant and NGO representative (Interviewee 10, 2004) agreed with this statement, asserting that 'African countries are partners to the EU, AMCOW is a partner, the government of Uganda is a partner, we are stakeholders.' A Secretariat member (Interviewee 7, 2004) illustrated this inequality of treatment: 'Everybody should have the same level of information . . . [but] most of the time Member States have more information than everybody else.'

Clearly, then, there existed a discrepancy between EUWI's partnership aspirations and the reality. A representative from a private water operator (Interviewee 8, 2004) claimed that ever since June 2003 stakeholders were asking for clarification on:

> who is a partner, who is a stakeholder, who is an observer, who is whatever, and what are the roles and responsibilities of those actors? What are the terms of reference of the Working Groups? What are the principles they must work to, for example stakeholder involvement? How do you treat documents? When do you send them out? What do you do with the comments on them? How do you ensure that the process keeps rolling?

It could be argued that this lack of equal treatment is attributable less to EUWI's structure, than to a lack of commitment on the part of stakeholders. An ex-Commission EUWI co-ordinator (Interviewee 5, 2004) suggested that 'the NGOs . . . are far too passive in the whole process . . . The private sector is the same, they are far too passive.' A Member State representative (Interviewee 12, 2005) pointed out that at the beginning of 2004, 'the list of people invited to the Multi-Stakeholder Forum had more than 200 people on it and only about 40 of them would come,' suggesting that this was because only those participants who were 'seriously interested' were still attending. However, it could be that stakeholder commitment to EUWI waned because EUWI failed to make stakeholders feel as if they were partners. José Frade (2004) of the

European Investment Bank (EIB) suggested that 'what is really important is to create a win-win situation where each stakeholder sees some gain and that the actions, objectives, and activities of that stakeholder would benefit from the initiative; that is really the platform that has to be found . . . and we have not yet fully reached that stage.'

However, it could also be argued that the inequality has been caused by structural tension within the EU Commission. The Environment Directorate-General (DG) and the Development DG, are the two main Commission DGs involved in the initiative, heading up the EUWI Secretariat (although the DG for External Relations, the DG for Aid Co-operation and the DG for Research and Technological Development are also involved). One Member State representative (Interviewee 13, 2005) reported that DG Environment and DG Development 'are quite different in the way they operate, the way they think, in what their focus is.' Le Génissel (2004) of France, stated that the two departments do not 'have the same philosophy and . . . it is difficult to see a unified point of view from the Commission' at times. A private sector participant (Interviewee 8, 2004) suggested that before EUWI was launched, 'DG Environment was of the opinion that we needed to change the way we do business, in particular on water' because previously there was no co-ordination: 'The Member States did their stuff, the EU did their stuff, but nobody really co-ordinated throughout.' DG Environment was apparently much more in favour of opening up the EUWI process to stakeholders than was DG Development, because, as a private sector participant (Interviewee 8, 2004) suggested, the development side 'weren't happy having someone interfering in their business.' This private sector participant (Interviewee 8, 2004) observed that the involvement of civil servants in DG Development 'with stakeholders has been near to zero,' which meant they had to change the way they did business, which was 'a huge task'. An independent consultant and NGO representative (Interviewee 10, 2004) concurred, stating that, 'There is a certain mindset within development people . . . it's not an area which is open to different ways of working.' An ex-Commission EUWI co-ordinator (Interviewee 5, 2004) explained that 'The donor community is used to working much more in a closed shop than for example in typical environmental policy making,' and was thus not experienced in engaging with non-governmental stakeholders to the same degree as were environmental policy makers and practitioners.

The new operational structure adopted in July 2004 tried to address the above concerns, in that, whilst governments remained predominant, other stakeholders now play a more significant role than before, most notably in the new multi-stakeholder SG, which is the driving force of the initiative, with extensive decision-making powers. Although high-level political and financial decision-making powers remain with the Council of Ministers, the multi-stakeholder SG is responsible for 'guiding the operational delivery of the strategic objectives and annual work programme of the Initiative' (Walshe *et al.* 2004). Thus, 'all decisions relating to the development of the Initiative for which Council involvement is not required at a horizontal level should be taken by the EUWI SG' (Walshe *et al.* 2004). The advent of the SG also helps to overcome the logistical problems involved in the representation of such a large number of members' views in the MSF (Interviewee 8, 2004). For Jacques Labre (2004) of Suez, 'the Multi-Stakeholder Forum was a very big meeting with a loose list of participants and it was not very clearly managed,' by contrast to the SG, which a Secretariat member (Interviewee 7, 2004) described as 'a kind of reduced Multi-Stakeholder Forum' with stakeholder representation, and at 'the heart of this initiative', adding that the Secretariat was no longer expected to head the initiative: 'the Secretariat is only there to facilitate the process . . . but not to lead.' As Frade (2004) of the EIB stated, 'the Secretariat in this process has of course to provide support but it cannot drive the process.'

Nevertheless, speaking in November 2004, Labre (2004) of Suez claimed that the Commission Secretariat was still very much in the driving seat, stating that even with the advent of the new organizational modalities, 'of course they lead . . . there is no doubt that the Commission leads.' The role of the MSF also continues to be a matter of some contention. One Member State representative (Interviewee 13, 2005) claimed that the MSF was still 'just a town hall meeting, a talk-shop where nothing is decided' other than who would represent the various stakeholder groupings on the SG.

It remains to be seen, therefore, whether the new organizational modalities will facilitate a multi-stakeholder *partnership*. As Annette van Edig (2004), a representative from the German Federal Ministry of Economic Cooperation and Development, said when interviewed in November 2004, 'we [the Steering Group] have met once so far . . . [and] it was a bit like a Multi-Stakeholder Forum meeting . . . we didn't have any real discussions.' A Member State representative (Interviewee 12, 2005) stated in April 2005 that the Steering Group 'is still a bit weak'. Labre (2004) of Suez held that the new organizational modalities could not overcome the deep-rooted institutional tension within EUWI, because 'the major role is, and will remain with the Member States . . . by no means can the NGOs and private companies decide anything in terms of financing or legal decisions. It would be against the institutional rule of the EU.' One thing, then, seems clear – that the paymaster partners will always have the last word in EUWI.

The EU-Africa Context

In this section, moving from partnership processes in the northern context to partnership processes at the regional level, we examine the establishment of partnership relationships between the EU and Africa, and then focus on the EUWI's Africa Water Supply and Sanitation Working Group.

The Establishment of an EU-Africa Partnership

The EUWI asserted the need for a bottom-up or demand-driven approach to aid, and one Secretariat member (Interviewee 6, 2004) suggested that this was how EUWI itself was initiated: 'it started as a response to a call from the south, and the north organized itself to respond to it. Since then, the partnership, or the collaboration we have between the European Commission and the Member States and the African Ministers' Council on Water has been well balanced.' An ex-Commission official and the former EUWI co-ordinator (Interviewee 5, 2004) explained how:

> before Jo'burg [and the WSSD] we had major political consultations . . . with Africa, with the African Water Ministries . . . we went to all their meetings and we discussed with them priorities and so on. And then we prepared a strategic partnership to be signed at Johannesburg on a head of state level. It was agreed by all African and European Union countries, with involvement also from stakeholders.

However, a Member State representative (Interviewee 13, 2005) disputed this interpretation of events, claiming that 'The African's didn't really get on board until Johannesburg. It was like an offer to Africa – "are you buying in to it?" . . . We signed the agreement but until then they hadn't really been involved.'

At the outset, EUWI focused on partnership between central governments, that is, between the EU Commission and Member States on the one hand, and AMCOW and African governments on the other. At Johannesburg, the *African-European Union Strategic Partnership on Water Affairs and Sanitation* was signed on behalf of Africa by the President of South Africa and the President of the Federal Republic of Nigeria, while on behalf of the EU it was endorsed by the President of the European Council and the President of the European Commission. Moreover, *The EU Water Initiative: Africa – EU Strategic Partnership on Water Affairs and Sanitation. Outline Strategy and 2004–2005 Work Programme* stated that, 'The EU relates directly to the African Ministers' Council on Water . . . a continent-wide group of African Ministers' established 'in 2002 to facilitate action by African governments in water resources development and management' (AMCOW-TAC & EU 2003, 22). Hence one Secretariat member's (Interviewee 7, 2004) comment that, 'On the political side we can talk about a north/south partnership.' According to Interviewee 5 (2004), genuine partnership relations were established between the EU and African governments in the initiation phases of the EUWI, and African governments 'had their say in the development of it in the same way as the European partners.'

Nevertheless, EUWI has been accused of being northern-dominated (Interviewee 9, 2004; Interviewee 10, 2004; Saywell 2004). Morley (2004) claimed in March 2004 that, 'it is very much a northern initiative' at the moment. A private sector participant (Interviewee 8, 2004) questioned the extent to which a genuine partnership could exist when 'one side has the problems and the other side has the money.' A Secretariat member (Interviewee 11, 2004) admitted that 'The difficulty in Africa is in having the capacity, financial and human, to provide the level of input that is necessary . . . So there is a lot that is happening and needs to happen to give AMCOW the capacity it needs so that it can partner in a much stronger way with us and with its other partners.' The charge of northern domination has plagued the partnership from the start, and it has not gone away: in February 2006 at an EUWI review workshop in Uganda, Reginald Tekateka from the Department of Water

Affairs and Forestry, South Africa, commented 'that there was a common perception among Africans that AMCOW was not a participant on the EUWI as it was EU driven and input from the African side was limited' (Mushauri 2006, 10).

Officially, according to *The EU Water Initiative: Africa – EU Strategic Partnership on Water Affairs and Sanitation. Outline Strategy and 2004–2005 Work Programme*, EUWI was 'based on the establishment of strategic partnerships in specific regions that draw together government, civil society, private sector and other stakeholders to help achieve water-related goals' (AMCOW-TAC & EU 2003, 5). Thus the partnership was committed to reach out beyond the governmental realm to include other stakeholders. However, in terms of the Africa-EU agreement made at the WSSD, an ex-senior Commission co-ordinator of the EUWI (Interviewee 5, 2004) noted that, 'The agreement was put out for comment to stakeholders but it was never put to them to sign.' He argued that EUWI 'is not a typical Type Two partnership . . . It is something between a Type One and Type Two, in a certain respect it is more like a Type One because it is really a government agreement.' The implication here is that while EUWI was envisaged as a platform for partnership between various stakeholders from Africa, the EU, and other international actors, in practice it developed as a partnership at the governmental level with some stakeholder participation. Indeed, one independent consultant (Interviewee 10, 2004) who was involved in the initiative from the outset, maintained that 'It was never really promoted as a Type Two by the EU . . . that wasn't the motivation behind it, to call it a Type Two, to get it on an approved list . . . that wasn't the game at all.' Instead, he suggested that the principal motivation behind the initiative was that because 'the political profile of water was increasing' in the run up to the WSSD, 'the EU wanted something to take to Johannesburg,' and thus EUWI was born.

In the next section, we examine EUWI's Africa WSS Working Group, to investigate for ourselves how far stakeholders other than governments have been engaged in partnership working.

The Africa Water Supply and Sanitation (WSS) Working Group

The overarching vision of the WSS WG, according to the initial Terms of Reference (quoted in EUWI 2004c, 3), was that: 'A strategic water and sanitation partnership is established between Africa and the EU that leads, in harmony with other efforts, to the achievement of the MDG both through attainment of physical targets and through improvement in the necessary underlying conditions of good governance, coordination and sector capacity.' Although the EUWI SG was given responsibility for driving the initiative as a whole and reviewing the work programmes and progress reports of the Working Groups, 'Decisions relating to the practical development of the Initiative at regional level lie with the Working Groups' (Walshe *et al.* 2004).

A Member State representative (Interviewee 13, 2005) involved in the WSS Working Group reported that it was a 'technical committee, representing the five sub-regions, east, west, central, south, and north Africa . . . It was a 15 member committee with three members from each sub-region.' This was claimed by a Commission official and member of the Secretariat (Interviewee 4, 2004) to be proof positive that, 'With respect to the government partnership – there we can say that we have achieved a proper north/south partnership because . . . AMCOW and NEPAD are represented in the meetings and are actively involved in discussions' within the Working Group. Another Secretariat member (Interviewee 11, 2004) pointed out that by November 2004, the Working Group was structured in such a manner that there were two co-chairs, 'There is a lead Member State . . . [and] a representative nominated by the Technical Advisory Committee of AMCOW.'

However, despite the claims that a genuine partnership had emerged between Africa and the EU at the governmental level, questions were raised about the way that this relationship was actually working within the WG. The minutes of the second meeting of the African WSS WG stated that in August 2003, 'AMCOW had expressed concern regarding the absence of African members in the various working Groups . . . [and] African representatives had not been invited to the WSS Africa WG meeting now taking place' (EUWI 2003a). To remove this imbalance, it was agreed that more of the future meetings of the WG would be held in Africa, making African governmental participation easier logistically and financially. But in the minutes of the WSS WG meeting on the 17 January 2005, concern was again raised about a lack of African governmental participation in WG meetings (EUWI 2005b, 1).

Moreover, even if the partnership between European and African governments was in good shape, partnership relations between governmental actors and non-governmental stakeholders, both northern and southern, were far from healthy. For example, a private sector representative (Interviewee 8, 2004) argued in March 2004 that:

> the Danes who are leading on the Water Supply and Sanitation group . . . see the EU as a partner, the African governments are partners . . . [but] there is no real role for stakeholders in the north and it's up to the Africans if they want to have local stakeholders involved, but they don't have a culture of doing that so normally they don't.

But a Secretariat member (Interviewee 11, 2004) asserted that 'at the Working Group level the European NGOs have . . . two complementary roles: one is to help develop the work programme and its focus and agenda . . . But the other role . . . is to bring in partner country or regional civil society organizations and allow them to play a big role as well.' This interviewee (11, 2004) also suggested that a similar role should be played by private sector participants, in that 'they are bringing knowledge . . . around sector planning . . . around management issues, and financing issues' but they could also prove useful in 'bringing in counterpart organizations.'

In terms of *northern* non-governmental stakeholder participation in the Africa WSS WG, it was weaker than expected from a multi-stakeholder partnership, especially in the early days of EUWI. A Commission official and member of the Secretariat (Interviewee 6, 2004) reported in April 2004 that the focus 'has been to develop the real partnership with the African partners [i.e. governmental partners] so maybe the stakeholders have been a bit lost.' Another Secretariat member (Interviewee 4, 2004) admitted at the end of March 2004 that 'stakeholder partners have not been involved [in the WSS WG] as much as one would have expected them to be so far.' An NGO representative (Interviewee 10, 2004) reported that 'The very first meeting [of the WSS WG in November 2002] . . . excluded everyone except Member States,' but after the following MSF, the WG led by Denmark was told by the Commission and other Member States, as well as non-governmental stakeholders, that it must 'invite a wider range of participants.' Accordingly, they invited Jacques Labre of the Suez Group, and Stephen Turner of WaterAid to the next meeting of the group, held in August 2003. However, because they 'were not selected by the NGOs and industry to represent them,' Labre and Turner declared that they had no mandate to represent these stakeholder groups' views as a whole (Interviewee 10, 2004).

It has been suggested by a Commission official (Interviewee 4, 2004) that:

> One of the difficulties may have been that the business as usual approach that has been followed in the water sector in the past was clearly not multi-stakeholder . . . In the old way of working, multi-stakeholder processes were not privileged and it was mostly a process based on dealing with government partners and then consulting stakeholders, but not truly involving them in the process.

This non-participative culture of the water sector would explain why non-governmental stakeholder participation in the WSS Working Group met with some 'resistance or reluctance' (Interviewee 4, 2004). However, another Secretariat member (Interviewee 6, 2004) maintained that it was 'difficulty more than reluctance', although he conceded that 'there was a bit of fear in the initial stages about involving too many participants, making the process completely uncontrollable.' Likewise, Frade (2004) of the EIB argued that if 'there is a lack of transparency . . . I don't think it is deliberate, it is just because there are still difficulties in feeling the way forward' and in co-ordinating so many different stakeholders. A former Commission official (Interviewee 5, 2004) referred to the caution expressed by the Member States involved in the WSS WG, stating that they 'don't see what the benefit is of having stakeholders involved . . . they are scared about what it means and what they will bring.'

A Commission official and Secretariat member (Interviewee 7, 2004) suggested that the reason given by Denmark for the exclusion of non-governmental stakeholders from WG meetings was that 'it was first necessary to have political agreement between governments on the EU and African side.' Later, 'in the second phase', Denmark claimed that stakeholders would be brought on board. A Member State representative from the WSS group (Interviewee 13, 2005) similarly argued that, 'at the first meeting we did not invite the private sector and NGOs . . . But we thought we had to get our internal act together and then expand.' However, a private sector participant (Interviewee 8, 2004) interviewed in March 2004, maintained that 'it is very much about individuals when it comes down to it, if the individual chairing a Working Group doesn't believe in stakeholder involvement then it is just impossible.'

This ambiguity related to the controversy over EU Member States making decisions without consulting non-governmental stakeholders in the MSF. At the ninth MSF in November 2003, it was accepted that 'some decisions could only be taken by governments' at the Working Group level (EUWI 2003b, 2). However, it was also agreed that 'where, possible, stakeholders should be involved in the process' (EUWI 2003b, 2). Despite this agreement, however, a representative of a private water operator (Interviewee 8, 2004) claimed that:

> the Africa Water Supply and Sanitation Working Group, led by Denmark . . . made a work plan which they didn't fully discuss with the Multi-Stakeholder Forum. They only gave a very short summary of it and said you can comment on the summary, and then without having the comments and without having the full discussion they went to Addis Ababa [in December 2003] to speak with the Africans who approved it and that's now a standing document of the EU Water initiative yet none of the stakeholders knew about it and knew what was really in there.

Nevertheless, there is no doubt of the good intentions of the MSF in wanting to encourage northern non-governmental stakeholder participation. In the minutes of the tenth MSF, held in March 2004, it was noted that 'Representatives of civil society and private sector underlined the lack of stakeholders representation especially in the WSS WG, and the need to open more widely the platform of discussion' (EUWI 2004a, 5). According to a Secretariat member (Interviewee 4, 2004), agreement was reached at this meeting that 'Working Groups should be multi-stakeholder in all cases.' Another member of the EUWI Secretariat (Interviewee 7, 2004) declared that the Commission wanted the Africa WSS Working Group 'to be open and more transparent in its way of working' and that 'on the Commission side we say that it is very, very important to have this multi-stakeholder character in each Working Group and in each meeting' so we 'hope very much that our Danish colleagues . . . will take this into account in the future.' Indeed, when the new organizational framework was adopted in July 2004, it clarified the criteria for the membership of each Working Group, and stipulated that: 'all working groups should comprise representatives from interested Member States, partner countries, the European Commission, operators and NGOs,' making the multi-stakeholder nature of the Working Groups more-or-less mandatory (Walshe *et al.* 2004). This led one Secretariat member (Interviewee 11, 2004) to argue in November 2004 that, 'with the structure made much clearer now . . . I think we are seeing a much stronger involvement at the Working Group level of the NGOs, the civil society organizations, and hopefully of the water utilities.'

Furthermore, in November 2004 the Netherlands took over from Denmark the position of co-chair of the WSS Working Group, and the Dutch were much more open to engaging non-governmental stakeholders in the Working Group than the Danes had been (Interviewee 12, 2005). The thinking was that the Working Group would now be, as a Member State representative (Interviewee 13, 2005) put it, 'in principle quite open,' though with the proviso that 'only those who are going to be active partners may become members.' The organizational framework stated that 'Participation entails a responsibility to undertake work actively within the Working Group' (Walshe *et al.* 2004). In other words, as a Secretariat member (Interviewee 11, 2004) stated, 'Membership of the group is [now] open to anyone who is prepared to contribute to the work of the Working Group . . . people are . . . not going to sit [on the Working Group] just because they think it would be useful to sit and watch.' However, he added, 'That's in theory what should happen . . . What we suffer from unfortunately is that there are a large number of people who decide that they . . . want to observe . . . but we're working on that and . . . the intent is that the Working Groups are full of working people from all the stakeholders.' A Member State interviewee (13, 2005), commented that 'it is a difficult thing to manage because how active do you have to be to be a member? It is very relative and so there is still a discussion on that.' Another Member State representative (Interviewee 12, 2005) reported in April 2005 that the Netherlands had introduced a qualification test for participation in the WSS WG – all stakeholders had to 'indicate on one A4 page what they can contribute and what they intend to contribute,' although how these documents were to be judged was unclear.

In summary, although northern non-governmental stakeholder *engagement* in the WSS WG had improved, it is not clear that *partnership* had been achieved. For example, the fact that the minutes from the fourth meeting of the Steering Group in October 2005 recorded that 'It was . . . suggested that WG chairs should involve partners and stakeholders in the preparation of the 2006 Work Programme,' showed that non-governmental stakeholder participation in partnership processes still had to be lobbied for, even after all the progress that had been made (EUWI 2005a).

With regard to *southern* non-governmental stakeholder participation in the WSS Working Group, there is a worse story. Morley (2004) of Freshwater Action Network argued that 'if one of its principles is that it is a stakeholder partnership that will be responding to the demands of the poorest in Africa then it certainly needs to work with African civil society and not just with governments.' A member of the EUWI Secretariat interviewed in 2004 (Interviewee 4) admitted that 'with regards to the involvement of southern stakeholders there is a lot of progress to be made,' and that although 'Important steps have been made in the right direction . . . this doesn't mean that we've fully set in place a truly north/south partnership, especially if we think about southern non-governmental partners.' Morley (2004) reported in March 2004 that at the regional level, in terms of civil society participation, 'It is very much government-based at the moment, they've had meetings with civil society in Africa but they don't really have a role to play.' And in terms of indigenous private sector involvement, van Edig (2004) said that 'private industry [both from the north and the south] I can't see at all. I don't know what they are doing or what their role is.'

One reason given for the lack of southern non-governmental stakeholder participation is that because of the desire to create a demand-led initiative, the involvement of non-governmental stakeholders in Africa was the responsibility of African governments. Accordingly, at the regional level, 'AMCOW will decide how to handle this issue' (EUWI 2003b, 2). But, as a Member State representative in 2005 (Interviewee 13) noted, it had 'taken some time for them [AMCOW] to accept the idea of a multi-stakeholder process because they are government people and are not used to working in that way.' This led Morley (2004) to argue that 'the process on . . . [the African] side isn't at all democratic. It's imperfect on . . . [the EU] side, but it's much more closed and much more government-led in Africa.'

However, a former Commission official and co-ordinator of EUWI (Interviewee 5, 2004) defended AMCOW, contending that:

> The African Council of Ministers is very new, it was only formed in Johannesburg and they have to get operational . . . It is very new for them to work together with other governments, and then to work in such an environment with the European Union is, of course, also completely new. It is not like classical donor work and the next step is to get stakeholders involved, and that is very hard.

Given an independent consultant's (Interviewee 10, 2004) comment that 'AMCOW took a long time getting its act together . . . In the first couple of meetings of AMCOW they didn't have proper agendas and they didn't have proper minutes, you couldn't work out what was happening,' it is hardly surprising that engaging non-governmental stakeholders was not a priority for AMCOW.

Another reason for the lack of southern non-governmental stakeholder participation was the growing recognition in development circles that the political commitment of southern governments was vital to secure sustainable reform in the water and sanitation sector. As one NGO interviewee (1, 2003) put it, if you by-pass African government by focusing directly on civil society, then you 'set up different levels and types of authority and responsibility,' and there is a danger that you 'set up a parallel system.' Another interviewee from a Kenyan NGO (Interviewee 9, 2004) believed that although in some cases you got more results 'when you overstep government processes and . . . work [directly] with civil society organizations or community organizations,' you 'can't ignore governments because at the end of the day they have to carry the process forward after any assistance comes.' Another explanatory factor was offered by a Secretariat member (Interviewee 4, 2004), who suggested that funding was an issue: 'the involvement of stakeholders from the south will need to be supported financially and the question remains open as to who should finance this and how we should organize this process.'

It was hoped that the formation of the African Civil Society Network on Water and Sanitation (ANEW) in October 2003 would improve the situation. An ANEW representative (Interviewee 9, 2004) reported that ANEW was formed as a way of 'channelling the diverse views and voices of people on how they thought Africa needed to move forward,' so that African NGOs could have 'a stronger reach' and be able to 'actually engage in these big decision-making processes [such as those of AMCOW, NEPAD, the EUWI and the Water and Sanitation Programme (WSP)] and try to shape them . . . engaging at the policy level.' This interviewee suggested that because decision-making processes now took place at a regional level and not just on a country-by-country basis, civil society groups in Africa 'are forced to be able to respond at that level.' She added that from an ANEW perspective:

> we really feel that we have some value to add and what we want to do from our position is make sure that this opportunity is not lost . . . you have a network which can reach out to different regions in Africa and there is a rapport between these different regions in Africa which has never happened before.

In December 2003 at the Pan-African Conference in Addis Ababa, 'First contacts between ANEW and AMCOW took place' (EUWI 2004a, 4). Speaking at CSD-12 in April 2004, an ANEW representative (Interviewee 9) argued that although AMCOW were 'not responsive to civil society voices' in the beginning, 'they are opening up – right now they ask for us more than they used to, they are discussing with us more than they used to.' Similarly, a Commission representative and Secretariat member (Interviewee 6, 2004) observed that at the outset 'there was reluctance from both the civil society side and from governments' to engage with each other, but since ANEW's inception, there had been an increased 'willingness to collaborate' by both African governments in the form of AMCOW, and the civil society groups represented in ANEW. Another EUWI Secretariat member (Interviewee 7, 2004) urged in early 2004 that 'we try to have them [ANEW] in the picture because we need them in this partnership.' Later that year, a third member of the EUWI Secretariat (Interviewee 11, 2004) stated that 'ANEW are becoming quite strongly involved with the Water Initiative which is very good . . . [and] AMCOW and AMCOW-TAC [AMCOW-Technical Advisory Committee] are having a debate on how they should engage with civil society . . . things are moving in the right direction . . . [but] it's been very hard.'

Although ANEW are not partners in EUWI (Interviewee 9, 2004), and there is no formal protocol or agreement for their engagement with AMCOW, at the fifth Ordinary Session of AMCOW, held in Nairobi in December 2004, it was agreed that the AMCOW President would 'consult with [a] Ugandan representative of ANEW to initiate [a] process for developing rules of procedure' (Schild 2004). At the Meeting of the Technical Advisory Committee and Executive Committee of AMCOW, held in Uganda in February 2006, a document was produced entitled, *CSO and Public Participation in the African Ministers Council on Water (AMCOW) Process: Guidelines for Engagement*. In this report (AMCOW 2006), the importance of engaging stakeholders was recognized for three reasons:

- External stakeholders have many different perspectives and experiences to offer in order to foster long-term, broad-based support for AMCOW's work.
- Engaging a wide range of stakeholders in addressing water and sanitation issues expands the reach and impact of strategies far beyond the capability of AMCOW's own limited financial and human resources.
- Active involvement of stakeholders at the national and local levels, where many problems needed to be addressed, and where many of existing programmes on water and sanitation should complement AMCOW's presence at the regional and global levels.

The report recommended that 'A strategy needs to be articulated to guide the interactions between AMCOW and civil society' with the following objectives:

> For AMCOW: To seek greater collaboration with civil society in the fulfilment of its mandate to ensure equitable access to water and sanitation in Africa. For civil society: To effectively network together with other stakeholders, build internal capacity through raising awareness of AMCOW's activities at national, regional and international level and opportunities to engage with AMCOW (AMCOW 2006).

Thus, CSOs' role should 'not be limited to that of observers,' and 'appropriate mechanisms that provide the required environment for optimal participation of all key stakeholders' needed to be developed (AMCOW 2006). The report recommended that a system of accreditation, together with a communication strategy aimed at outreach to civil society, should be developed; that AMCOW should institutionalize engagement with civil society in dialogue sessions with the Technical Advisory Committee and AMCOW Ministers; and that AMCOW should 'encourage sub-regional Ministerial and Technical Committees and individual governments to identify and engage CSOs in planning, implementing and monitoring programmes to achieve the goals and action plans developed by the Council' (AMCOW 2006). The report ended by noting that, 'For the process to be successful, AMCOW will have to dedicate specific resources and accommodate these costs in its annual budget, to facilitate this process of engagement' (AMCOW 2006). Therefore, although the institutionalization of civil society participation in AMCOW processes has been slow, some preparatory steps have been identified to speed it up. However, it is not clear whether any of these steps have been implemented.

The matter of financing participation affected not just civil society but governments as well. AMCOW participation in the Africa WSS Working Group meetings held in Europe was made possible largely by EU Member States and the European Commission financing the travel costs and expenses of their African counterparts, though 'Only one government, Sweden, has paid for African civil society members to attend EUWI meetings' (WaterAid & Tearfund 2005, 3). As one Member State representative (Interviewee 12, 2005) reported, ideas have shifted towards holding Working Group 'meetings more frequently in Africa' to encourage greater African participation, particularly the participation of other AMCOW-TAC members and those from the governments of the countries within which EUWI was working. When the EUWI Steering Group met in May 2005, it agreed that each Working Group must set down an annual budget as part of their yearly Work Programme, which should include budgeted costs for '(i) consultants, (ii) non-European partners' participation, and (iii) operating costs such as rentals, materials' (EUWI 2005c). Nonetheless, it was noted in the minutes of the joint meeting of the IWRM and WSS WGs in August 2005, that, 'In practice . . . many EU countries do not financially lend a hand to African partners' (EUWI 2005g). Therefore, in 2006, when the IWRM and WSS Working Groups merged into the Africa WG, the possibility of establishing a Trust Fund (made up of Member State contributions) for the Africa WG, to support its activities and fund African partners' participation in WG meetings was being explored, which would possibly be substituted by a grant contribution from the ACP-EU Water Facility (EUWI Africa Working Group 2006b). It is not clear if a Trust Fund has been put in place but a proposal was submitted to the Water Facility for support for the Africa WG in June 2007. It did not, however, receive positive initial feedback, although a revised submission was planned for August 2007 (EUWI Africa Working Group 2007).

A Secretariat member (Interviewee 6, 2004) suggested that southern non-governmental stakeholder participation in EUWI would increase when work was undertaken in 'specific countries and at the country level because as long as it remains a more global and theoretical process it is difficult to really involve everyone.' This leads us to the next section – the examination of the work of EUWI within a specific country, in this case, Zambia. Ten pilot countries were chosen by AMCOW in which to begin the country dialogue processes (Ghana, Cape Verde, Ethiopia, Rwanda, Mozambique, Zambia, Congo Brazzaville, Central African Republic, Egypt and Mauritania), with one Member State or the Commission taking the lead in each (EUWI Working Group on Water Supply and Sanitation in Africa 2005, 2). The country level actions were to have two phases: 'an initial phase (dialogue) and a follow up action phase' (EUWI 2004c, 13). Both the identification of the pilot countries by AMCOW, and the matching of interested Member States to these countries, took a long time to complete, for two reasons. First, many African countries were vying to be included as pilot countries as they wrongly believed this would mean increased funding. Second, there was a lack of commitment on the parts of some Member States to take the lead in particular African states (Interviewee 12, 2005; Interviewee 13, 2005). As one Member State representative (Interviewee 12, 2005) explained in April 2005, the initiative had been going for three years, 'and out of the ten [pilot] countries we still have four or five countries that we haven't even identified an EU Member State to take the lead on.' However, Zambia was the first country in which a country dialogue process was initiated, with Germany as the lead Member State, and it is to EUWI in the Zambian context that we now turn.

The Zambian Context

Background to the Country Dialogue Process

According to a Member State representative from the Africa WSS Working Group (Interviewee 13, 2005), country level dialogues should be inclusive, and were 'not something which should be limited to EU Member States, the Commission and then the country.' Instead they 'should be broad based . . . involving those donors who are outside the EU family . . . [as well as] the World Bank, the UN family, [and] the NGOs' and, presumably, the private sector. The September 2005 draft version of the *Strategy for Development of the EUWI* upheld EUWI's commitment to involving non-governmental stakeholders or CSOs in the country dialogues, stating that:

> Many partner country governments, at least in Africa, are lukewarm to involving CSOs in policy dialogue and see the private sector as a potential provider of goods and services. The EUWI should seek to safeguard the involvement of stakeholders in the policy dialogues and, as appropriate, make this a priority at country level (EUWI 2005e, 5).

The original idea behind the country level dialogues was that they would be split into two phases. The first phase would be a short six-month process whereby relevant parties were briefed on EUWI; an overview of the water and sanitation sector established for the country; and an action plan developed. The second phase would be implementation. As an EUWI Secretariat member (Interviewee 6, 2004), speaking in April 2004, explained, the idea was to:

> identify what is going on [within the country] and how the Water Initiative can add value to existing processes. If nothing is there to build on, to identify who the main actors are and start a dialogue with them, and identify their needs. The idea of the Water Initiative is not to replace or duplicate anything that already exists, it is to find where the approach of the initiative and the instruments that the partners of the Water Initiative are bringing with them could add value, or could fill gaps, or improve what is already going on.

However, when the Netherlands took over from Denmark as the lead EU Member State in the Africa WSS Working Group, the purpose of country dialogues was altered. A Member State representative (Interviewee 12, 2005) observed that under Denmark, 'the dialogue was being undertaken in terms of a policy dialogue, a national dialogue, in terms of how stakeholders were participating in the MDG process, financing of the investment programmes etc.' But 'it wasn't really very result orientated,' and there was a feeling that it needed to be made 'more output orientated . . . not just talking for the sake of talking, but talking as a way of getting things done in the field' (Interviewee 12, 2005). In a document produced by the EUWI Working Group on Water Supply and Sanitation in Africa (2004) entitled, *Proposal to Accelerate Progress on National Policy Dialogues*, it was suggested that the reasons for lack of progress in the country dialogues included the following:

- the purpose of the dialogues was not well understood by all stakeholders involved,
- there was a lack of communication on the objectives of the EUWI and the country dialogues with the selected countries,
- there was no prescribed format or blueprint for how the dialogues would be conducted,
- it proved cumbersome to bring the parties together in the ten countries selected,
- the dialogues could become time consuming both for the governments involved and for the EU MS missions, and
- the dialogues were not seen as action-oriented, and the outcomes envisaged (value added) were not clear.

The new thinking was that, as a Member State interviewee (12, 2005) suggested, country dialogues should bring:

> all the stakeholders in water and sanitation together . . . [in a specific country] and with the help of some consultants or facilitating bodies like WSP, try to develop a clear roadmap . . . which indicates the way the country intends to meet the MDGs target 7 for water and sanitation . . . A strategy to reach [un-served people] . . . in terms of capacity needed to reach the goals . . . [and] a clear MDG investment plan.

Once a long-term strategy and investment plan was developed, with the help of the EUWI Finance Working Group, it was anticipated that 'EU Member States may call a round-table to get donors and the private sector, and whoever is willing to participate, to agree on their contributions' (Interviewee 12, 2005). Thus the country dialogue process would now focus on 'agreeing on targets and outputs to be reached every year, and linking that to the inputs that are required,' with 'mechanisms in place that allow the review of the situation every year' (Interviewee 12, 2005).

Mozambique was to be the new testing ground for this approach, but country dialogue processes had already begun in Ghana and Zambia. Indeed, the majority of the work in Zambia took place before any change in approach was adopted by the Working Group. The implication was that the country dialogue process, which began in Zambia in May 2004, lacked focus and was not results-orientated. As a Member State representative (Interviewee 12, 2005) asserted, 'there were meetings in Ghana and Zambia but they were inconclusive and it has not yet developed into an action oriented strategy.' The country dialogue experience in Zambia probably played a large part in these new ideas on how country dialogues should be undertaken. In a document entitled, *Summary and Lessons Learned from the initial steps for starting the Country Dialogue in Zambia* (A-EUWI 2005, 1), it was suggested that the process in Zambia 'is a pilot exercise for preparing and developing an appropriate approach for the Country Dialogues in another nine African countries.' One of the conclusions of this report was that in future:

> Before starting a Country Dialogue, a clear framework of a long-term strategy in the country should already [have] been envisaged (e.g. 10-years work program to reach the water related MDGs . . .) and the respective financial resources to facilitate this should be mobilised for budgeting. It should be seen as a long-term commitment instead of an ad-hoc action (A-EUWI 2005, 4).

This was a recognition that 'it would have been good to have a longer-term strategy to discuss with the participants instead of leaving the process completely open' (A-EUWI 2005, 5).

Thus, while a Commission official and member of the EUWI Secretariat (Interviewee 11, 2004) reported that future dialogues processes were to be done differently, with new terms of reference, and that the Zambian dialogue would need to be 'realigned' with them, this change in approach was, at least in part, due to the knowledge that had been gained through the Zambian experience..

Overview of the Country Dialogue Process in Zambia

In Zambia, the major problem was less the supply of water than making it and sanitation available to people: 'Zambia has ample water resources . . . [but the] main challenge, apart from a few areas of the country, is access to water and sanitation services' (EUWI 2004b, 13). According to an EUWI Working Group on Water Supply and Sanitation in Africa (2004) document, when the Zambian country dialogue process was begun, the aims of the country dialogues were, in the immediate future, to:

- raise awareness about the EUWI,
- identify policy issues and institutional bottlenecks that impede investments in WSS, and
- set up a participatory approach to define country actions consistent with the MDGs and the WSSD targets.

In the long term, the dialogues were designed to contribute to:

- harmonization of donor procedures,
- joint donor-supported programmes for capacity-building,
- piloting of innovative financing mechanisms, and
- a better socio-economic justification for increased spending on water sector development.

Since the early 1990s, the lead donor in the Zambian water sector was Germany (along with Ireland in the sub-sector of rural water supply and sanitation (RWSS)), thus it seemed rational that they would take the lead on the EUWI country dialogue process in Zambia (EUWI 2004b, 14). The first phase of the Zambia country dialogue was carried out in May 2004 when Germany funded two consultants (one external – Eric Buhl-Nielsen from Denmark, and one internal – Moffat Mwanza from Zambia) for 'a baseline study of the water sector in Zambia' (Interviewee

13, 2005), to see 'what the situation was previously, what mechanisms were put in place to sort out that situation, how we were going to meet the Millennium Development Goals, how the sector could move forwards, and what was being done to try and do that' (Mwanza 2005). The consultants completed a study on the status of the water sector (the *Zambia Sector Status Report* (EUWI 2004b)), and a proposed strategy to take EUWI forward in-country (the *Zambia Country Dialogue Process Report* (EUWI 2004e)), both of which were prepared using desk-based resources supplemented by data collected from interviews with the 'cabinet office, ministries involved in water affairs, local governments, water institutions, the regulator, water and sanitation utilities, donors including the World Bank, [and] NGOs' (A-EUWI 2005, 2).

A workshop took place in Lusaka, the Zambian capital, in May 2004 to increase understanding of EUWI, review with stakeholders the background paper and its findings, and agree upon a proposed way forward for the second phase of the country dialogue in Zambia. Mwanza (2005) said that the workshop sought to 'critically [examine] . . . what advantage or added value the Africa-EU Water Initiative [A-EUWI] would bring to the sector – the comparative advantage.' It was not the intention that EUWI should 'duplicate ongoing activities,' instead it sought to identify gaps within the sector which it could play a role in filling (EUWI 2004e, 4). The *Zambia Country Dialogue Sector Status Report* (EUWI 2004b, v) listed four potential action areas where it was thought EUWI could prove helpful:

1) Assist the SAG [Sector Advisory Group], PRSP [Poverty Reduction Strategy Paper] water working group with generic arguments that can provide a compelling argument for increasing the prioritisation in the WSS [sector].
2) Assist in the resolution of the institutional [inter-ministerial] conflict by providing an independent, neutral, facilitating and arbitration support.
3) Support the development of an [sic] periodic joint Zambian-donor value for money study to lay the basis for future SWAP [Sector Wide Approach] and harmonised funding strategies.
4) Assist the development of the local private sector and the advocacy role of civil society.

The reasons why EUWI had a comparative advantage in these four action areas were explained, respectively, thus (EUWI 2004b, v):

1) Prioritisation: Prioritisation comes down to political will. The Johannesburg declaration [of the EUWI] signed by the 4 presidents and backed by the key findings of the WSSD (more attention to sanitation and water) is the highest expression of political will available and is potentially a unique prioritisation tool of the Africa-EU working in partnership with NEPAD, [and] SADC [Southern African Development Community].
2) Institutional Conflict: The Africa-EUWI brings neutrality that other donors and Zambian parties may not have. The Africa-EUWI does not have a past and will not establish itself as 'a player' in the future.
3) Value for Money in RWSS: Value for money requires benchmarking against best practice within the country and within the region. The Africa-EU WI represents a partnership between African and EU that invests more than 1 Billion € per year in the water sector in Africa. This partnership working with already established regional initiatives (e.g. WSP, NEPAD, SADC etc) has the potential to contribute not only to the analysis but more importantly to the dissemination and implementation of the consequences not only within Zambia but regionally.
4) Private sector and advocacy in civil society: A central insight from the very start of the Africa-EU WI was that the challenges of meeting the MDGs for water and sanitation required a joint government, civil society and private sector approach. The multi-stakeholder approach of Africa-EU WI could help to mobilize not only government partnership but partnerships that extend to the private sector and civil society.

At the national Zambia-EUWI workshop, 'a list of actions for follow up of the country dialogue were agreed upon . . . One of the main actions being to continue the country dialogue until an action plan was reached and the result disseminated at a regional workshop' (EUWI 2004e, 4). However, phase two of the country dialogue process – the action phase – failed to materialize, and the process stalled. As Peter Sievers (2005), the DANIDA (Danish International Development Agency)

Programme Co-ordinator for the Water Sector in Zambia observed, 'the EU Water Initiative itself is really a bit dormant.' An international NGO representative (Interviewee 17) stated in May 2005 that 'very little discussion takes place these days on the European Union Water Initiative.' In the same month, a World Bank and WSP representative in Zambia (Interviewee 22, 2005) maintained that 'there is an EU Water Initiative but I'm not sure that anybody knows what it's doing . . . If it has taken off, I haven't been involved in it.' Mwanza (2005), a Zambian consultant who helped to start the country dialogue process in Zambia, offered two reasons for this lack of progress: first was the lack of an in-country anchorage for the initiative; and second was ambiguity over the 'specific, tangible things that the European Union was going to bring . . . [the thing that] has held it back the most is the clarity on what should be done and what the next step is' (Mwanza 2005). Let us look at these two explanations in more detail.

In-Country Anchorage

The first issue raised by Mwanza – lack of in-country anchorage – reflected a feeling among some in-country partners in Zambia that EUWI lacked local ownership. Dr. Zebediah Phiri (2005), formerly of the Water Resources Action Programme (WRAP) (a governmental programme supported by various donors aimed at supporting the National Water Policy in developing an IWRM framework) in Zambia, said that:

> I see it more as top-down – something is decided . . . in Brussels and people say 'lets go and support Africa' . . . then after that has been decided that's when all the countries are asked to try and fit into that vision . . . [but what] we should be saying [is] that this is a Zambian project, that way it is sustainable, otherwise when the donors leave it ends . . . if it was more bottom-up [to begin with] they might have been able to realize what would work and what would not work, but the way it is, things have already been agreed and forms have been designed, a website created . . . [and then they realize that their pre-designed approach] won't work, but time is lost.

Similarly, Charles Chisanga (2005), of the Water and Sanitation Association of Zambia (WASAZA), held that 'things are imposed, they are not really of our own making.' Likewise, a Development Cooperation Ireland (DCI) representative in Zambia (Interviewee 16, 2005) stated that 'I think it is experienced as a top-down approach rather than a felt need here . . . It's external and there hasn't been much of a buy-in from the local stakeholders.'

The initiative did try to find an in-country anchor within the Zambia government to 'take ownership of the process' (Werchota 2005). The *Zambia Country Dialogue Process Report* stated that 'A pre-condition for support to the 2nd phase of the country dialogue is the setting up of an institutional anchorage for the country dialogue in Zambia' (EUWI 2004e, 9). Roland Werchota (2005) of GTZ (German Technical Cooperation, the coordinating body for implementing EUWI in Zambia) explained that that this meant 'that we get someone who is responsible for the initiative who will then start working with AMCOW.' However, this was not an easy task, and EUWI struggled to find an institutional home within the Zambian government. The major problem was that in Zambia, the responsibility for the water supply and sanitation sector was shared between five ministries: the Ministry of Finance and National Planning (MoFNP), the Ministry of Local Government and Housing (MLGH), the Ministry of Tourism, Environment and Natural Resources, the Ministry of Health, and the Ministry of Energy and Water Development (MEWD) (EUWI 2004b, 3). Dr Phiri (2005) noted that coordination between these ministries was less than perfect: often 'the government doesn't seem to agree about what needs to be done . . . [and] what the priorities are.'

Moreover, there was deep-rooted conflict between the two principal ministries – the MEWD and the MLGH. The MEWD had 'responsibility for water resources management,' whereas the MLGH had 'prime responsibility for Water Supply and Sanitation infrastructure planning and resource mobilization' (EUWI 2004b, 3). Rivalry was in part because of competition for resources between the two institutions, 'especially for large-scale financial resources for implementation' (EUWI 2004b, 17). But it was also caused by historical disputes about which ministry was responsible for water supply implementation. As Sievers (2005) explained, the major problem was that the MEWD did not 'want to relinquish the implementation role [it previously held] to local government,' even though it had been legally established as a local government function over ten years ago. Indeed, local governments often did not have the capacity to fulfil all their water supply and sanitation obligations, and still needed the MEWD to continue in an implementing role (Kapotwe 2005; Lupunga 2005; WaterAid 2001, 3). Werchota (2005) pointed out that, 'although the legislation is clear and very good, there are different interpretations of the legislation, and there are also structures on the ground which respond to the old system.' A WSP and World Bank representative (Interviewee 22, 2005) noted that 'The issue is a historical issue . . . there are issues of power, of history, it's a whole range of issues that relate to the way the sector has evolved.' But a MEWD official (Interviewee 19, 2005) claimed that 'there is no institutional conflict, it is about people,' and Paul Lupunga (2005) from the MoFNP stated that 'You can never divorce personalities from institutions and I think both sides do have strong personalities.'

Whatever the reason for these disputes, WaterAid (2001, 3) claimed that 'The upshot of this upper level dispute is continuing uncertainty and inertia at District level . . . [which] ultimately translates into levels and standards of service provision in urban, peri-urban and rural areas that remain below acceptable standards.' The EUWI *Zambian Sector Status Report* (EUWI 2004b, 16–17) noted that the ongoing conflict between the two ministries over leadership of the sector led to:

- Two working groups updating the water sector PRSP instead of one.
- Two ministers attending the Pan African Conference in Addis Ababa (Dec. 2003) instead of one.
- Dual implementation of borehole programmes by DWA [Department of Water Affairs] and Local Authorities.
- An inability to establish sector overview on present coverage and investment requirements.
- A confusion and resultant varying practice amongst donors one [sic] which organisation to support for RWSS.
- Donors perceived and encouraged to be taking sides on issues that are internal Zambian concerns.
- Delays in the resolution of how to commercialise or make best use of government drilling rigs.
- Uneven attendance at sector workshops depending on which ministry is the host.
- Rejection by one party of the other's sector plans and strategies.

- Central government sending contradictory signals in terms of policy statements, cabinet memorandums, ministerial restructuring and staff establishment, budget allocations – depending on lobbying successes and failures.
- Differences in the strategy in how to serve peri-urban areas (differences not arising from healthy piloting of alternatives).
- The DTFs [Devolution Trust Fund's] fund raising credibility damaged due to differences in opinion on where it should be hosted.
- Good governance mechanisms in the UWSS [urban water supply and sanitation] [sector] cannot be agreed on thus hampering implementation.
- Significant donors threatening to withdraw from the sector.

The report (EUWI 2004b, iii–iv) pessimistically concluded that:

> The single most vital challenge facing the WSS sector today is conflict between the Ministry of Local Government and Housing (MLGH) and the Ministry of Energy and Water Development (MEWD) over sector leadership within the RWSS sector. As a result of this conflict the RWSS [rural water supply and sanitation] sector is close to being dysfunctional and the UWSS sector is seriously compromised: many sector initiatives are being duplicated and many more are not being done at all. A Sector Wide Approach to Planning is unlikely to emerge until this problem is resolved . . . Without a resolution of the inter-ministerial conflicts, the ongoing reform efforts will not be enough to remove the obstacles . . . to achieving the MDGs.

Despite these obvious tensions, however, the consultants working on the dialogue process proceeded to locate anchorage for the initiative in the MEWD. They felt that as 'water policy in this country is supposed to be implemented by the Ministry of Energy and Water Development . . . it [was] . . . obvious that this ministry should have the anchorage for the initiative,' and should provide leadership (Mwanza 2005). This decision was prompted by the fact that the minister who represented Zambia on AMCOW was the Minister of Energy and Water Development. As a result, the Minister for Local Government and Housing, although invited, chose not to attend the EUWI national workshop because she felt her ministry had been sidelined: 'it was a water initiative on water supply so they should be the lead ministry' (Mwanza 2005). An aid agency, DCI representative (Interviewee 16, 2005) supported MLGH, stating that:

> in the DCI view it is not located in the ministry where it should be located . . . It's about Millennium Development Goal achievement and it's about water and sanitation provision, and the Department for Infrastructure and Support Services is the implementing agency within the Ministry of Local Government and Housing that has that mandate.

On a more positive note, one donor representative (Interviewee 20, 2005) believed that 'it will end up as a proper partnership; like every new initiative or programme there are lots of teething problems and until those are sorted out you're going to get these bottlenecks.' Furthermore, Werchota (2005) declared in May 2005 that co-ordination between the MLGH and MEWD was improving, because 'at the last Sector Advisory Group meeting (which is part of the PRSP process) both the Minister from Local Government and the Minster from Energy opened it.' While this did not mean that the anchorage issue for EUWI had been resolved, it did suggest that the MEWD and the MLGH were beginning to overcome some of their mutual hostility.

However, Werchota (2005) believed that the fundamental solution to the anchorage problem was a stronger presence from AMCOW. In a *Lessons Learnt* document produced about the Zambian country dialogue process, it was acknowledged that local ownership had been a problem, and the document stated that 'In order to avoid that the Country Dialogue is regarded as a donor-driven initiative, it is necessary to invest more time in identifying a national "champion" – if possible from AMCOW-TAC' (A-EUWI 2005, 4). Peter Sievers (2005), interviewed in May 2005, argued that 'AMCOW have to appoint somebody here so that we can have a dialogue partner . . . AMCOW should also request regular reports on the progress of the country dialogue . . . [Then AMCOW] can help to solve institutional conflicts emerging between government institutions . . . [and] promote exchange.' As a Member State representative (Interviewee 12, 2005) noted, one of the problems was that 'AMCOW is not communicating well with its own countries.'

Although two members of AMCOW-TAC participated in the national workshop for the country dialogue process in Zambia in May 2004, (one from Kenya – Sylvester Kai – and one from Ghana – Emmanuel Nkrumah), which Mwanza (2005) claimed made it not 'just a European thing', AMCOW as an institution had very little involvement in, or contact with the Zambian country dialogue process and its participants. Werchota (2005) argued that:

> what's missing is cooperation and structure further down from AMCOW in Africa . . . [which] is needed to make sure that whatever dialogue takes place on the top level, between the EU in Brussels and AMCOW, is then coming down to the countries . . . AMCOW for us agencies here is still very much an institution which we get very little feedback on. The Zambian partners should deal with AMCOW . . . that's the missing link . . . because nobody from government here . . . is dealing with AMCOW . . . [As a result] it's always driven a little bit from Brussels . . . I think AMCOW has to do more.

These linkage deficiencies were due, in part, to the fact that AMCOW was a new organization with limited resources. Nonetheless, after phase one of the country dialogue process in Zambia, the need for AMCOW to have more involvement in country dialogues was widely recognized. A Member State representative (Interviewee 12, 2005) noted that in Mozambique, 'AMCOW-TAC is going to be much more closely involved . . . [and their representatives will] hold discussions with the Mozambique government . . . it's a joint exercise in which AMCOW is playing the same role as the EU, in that we are together talking to the government of Mozambique.'

AMCOW did not get involved in appointing a national anchor for EUWI in Zambia. Instead, at the May 2004 workshop it was agreed that institutional anchorage would be provided by the National Water Supply and Sanitation Council (NWASCO), the regulator of water supply and sanitation providers in Zambia, and that Osward Chanda, the Director, would take over this responsibility in the interim, until a permanent home could be found for the initiative within government. NWASCO 'supports the government to develop policies, set standards and guidelines, license Water and Sanitation utilities and monitor their performance. It also takes necessary action to provide for efficient and sustainable provision of services throughout the country' (EUWI 2004b, 3–4). At the EUWI May 2004 workshop, NWASCO was asked to take 'the anchorage issue forward . . . and ensure that the matter is decisively handled and followed up – ultimately at the Cabinet level' (EUWI 2004e, 10). However, eight months later it was reported in the minutes of a meeting of EU donors agencies in January 2005 that:

> the anchorage of the dialogue in Zambia has not progressed after the provisional nomination of a Zambian representative (Mr. O. Chanda, NWASCO) [thus] it was agreed that MoFNP [Ministry of Finance and National Planning] should be approached by the EC to initiate a meeting with the major players in the sector in order to obtain a final nomination.

The MoFNP was chosen 'because they are mandated by Cabinet to make these kinds of decisions' and therefore people 'have to listen to them' (Werchota 2005). Although 'The meeting was requested by the EU Commission in Zambia on behalf of the EU' it was Werchota of GTZ who met with a MoFNP official to discuss the matter (A-EUWI Zambia 2005b).

However, despite the fact that the minutes of a meeting between Werchota and Paul Lupunga (the official responsible for the German desk) concluded that 'In order to progress with the anchorage process, MoFNP could invite the mentioned stakeholders (MEWD, MLGH. NWASCO, EU Commission and member state representatives) in order to find an agreement on the way forward,' it appears that the anchorage issue was not discussed at all, but GTZ merely asked the MoFNP to make a decision on who was to be responsible within the Zambian government for EUWI (A-EUWI Zambia 2005b). Such top-down methods were unlikely to help the formation of constructive working relationships.

Furthermore, there was a call not only for the greater involvement of AMCOW, but also for a more prominent role to be played by the EU itself and their Zambian delegation. An international NGO representative (Interviewee 17, 2005) suggested that, 'Maybe if the EU itself had taken on the responsibility [for spearheading the EUWI in Zambia] things would have been somewhat different and there would have been a better outcome . . . I think the European Union itself should play a major role, and the delegation itself should be better placed in that.' Sievers (2005) explained that although the Commission argued that 'the Member States have to drive the process' in Zambia, 'when we work within the context of the EU Water Initiative we have the Commission [or rather the EU Commission's Delegation in Zambia] to chair, and we have their letterhead . . . so that we operate under the European flag.' Nonetheless, the EU Member States who were active in the water sector in Zambia felt that both the EU and AMCOW should be more heavily involved in the work of the EUWI in-country, to 'add political power at a higher level' (Sievers 2005). One donor agency representative (Interviewee 20, 2005) expressed disappointment at the EU's inertia: 'if the donors didn't take the initiative forward, nothing would happen . . . [because] very little is happening in the water sector as far as the [EU] delegation is concerned.' Brussels did not put any pressure on the EU delegation in Zambia to get involved in EUWI, so while the delegation chaired meetings, it did very little else, because it was reluctant to get heavily involved in the initiative as water is not its focal area; it is primarily concerned with the transport sector.

Another reason for the weak presence of EUWI on the ground in Zambia is the top-down nature of EUWI's approach. At a meeting in January 2005 of the EU donor agency representatives, including representatives from the European Commission (EC) delegation in Zambia, to 'Discuss the way forward for the country dialogue in Zambia,' it was noted that:

> The A-EUWI has started as a top down approach with the signing of the Johannesburg Declaration by two Heads of States (Nigeria and South Africa) and the European Council and Commission. Therefore, it is recommended that the relevant EU working groups as well as AMCOW take measures to support the initiation of the political will in each African country (A-EUWI Zambia 2005a).

Werchota (2005) claimed that this top-down approach was beneficial:

> the comparative advantage of the EU . . . is that it is high-level with a dialogue between two continents, with AMCOW on one

> side and the EU on the other side . . . what we need to do . . . as soon as we have the anchorage person, [is] have a working relationship or co-operation between the Zambian anchorage person and AMCOW.

Similarly, Sievers (2005) suggested that the EU Member States active in the Zambian water sector 'agreed that where it can really add value is at the higher level, at the continental level, the EU/AMCOW level.' However, the danger was that EU Member States would be reluctant to take responsibility for the in-country EUWI work. According to a DCI representative in Zambia (Interviewee 16, 2005), even the EU Member State delegations in Zambia 'experience it as external to us.'

Ambiguous Aims and Objectives, and the ACP-EU Water Facility

The second issue identified by Mwanza – lack of clarity surrounding the initiative's aims, objectives, purpose and focus – further complicated the situation in Zambia. Dr. Phiri (2005), formally of WRAP, when interviewed in May 2005, stated that when he learned about EUWI at the end of 2003:

> there was a recognition that there were already a lot of other initiatives [under way] so the EU people were simply saying . . . we just want to add value to what is already going on . . . but how that value was going to be added was not clear . . . There were those sort of questions being asked and I am not sure that they have been answered yet . . . I know they have . . . a long list – a shopping list – of areas for interventions . . . [but] the initiative needs to have some focus . . . it has to be clear what it is about . . . There were a lot of expectations when it was launched but all that is now waning.

Mwanza (2005) asked: is 'it to support water resource management and development, or . . . [is it] just about water supply and sanitation?' In other words, what was EUWI going to do in Zambia? If one of the consultants who built the foundations for the dialogue process did not know the answer to this question, then it was hardly surprising that Zambian stakeholders were confused, and that the two main ministries involved could not agree on who should provide the focal point for initiative within Zambia.

It became clear that the country dialogue processes had to address both water resource management and water supply and sanitation, because at the beginning of 2006, the two EUWI Working Groups on Africa, the WSS WG and the IWRM WG were combined, in an acknowledgement that any strategy to improve WSS must be part of an overall IWRM strategy. A Steering Group Discussion Paper on the topic of combining the WGs stated that 'There is general agreement . . . that the country dialogues . . . will be undertaken within the framework of an integrated approach to water management . . . IWRM and WSS must therefore feature together in the country dialogues, and treating them separately should not continue' (EUWI 2005f).

One reason for a lack of detailed objectives for EUWI in Zambia was because it was felt by Germany and the consultants used, that it was critical to 'anchor the country dialogue within Zambia before developing detailed recommendations' (EUWI 2004b, v). This was a "chicken and egg" problem: a lack of clarity over objectives made finding institutional anchorage for the initiative difficult, but at the same time it was unwise to create detailed recommendations without the initiative having an in-country champion, and thus a sense of ownership within Zambia. The former imperative prevailed, and in the final draft of the *Strategy for Development of the EUWI,* produced in September 2005 after the Zambian process had stalled, it stated that, 'Broad agreement and understanding of the objectives and means of the EUWI is the essential first step of building ownership by partner country governments and local stakeholders' (EUWI 2005e, 4).

Another reason for the controversy over the aims of EUWI in Zambia was its confusion with the EU Water Facility (WF). Many Zambian stakeholders saw the initiative and the country dialogue process as inextricably linked to the African, Caribbean and Pacific (ACP) European Union WF. Although EUWI was not a funding source, and offered no new financial aid, in 2004 the ACP-EU Water Facility was launched with €500 million – funds which were part of a conditional allocation to the ninth European Development Fund (EDF). The overall objective of the ACP-EU Water Facility was to: 'Contribute to poverty reduction and sustainable development through the achievement of the specific Millennium Development Goals (MDGs) and World Summit for Sustainable Development (WSSD) targets on water and sanitation in ACP countries,' and its specific aim was to 'Boost the sustainable delivery of water and sanitation infrastructure and improve water governance and Integrated Water Resources Management (IWRM) practices in ACP countries by helping to address the financing gap' (EuropeAid 2006). It was not intended that the Facility would fully fund projects, but rather that it would be used as a source of funds to catalyse and leverage finance from other actors.

When the possibility of a Water Facility was first mooted (and ever since) some critics urged that the Facility should be an extension, or the funding arm, of EUWI, with very strong connections between the two (Interviewee 8, 2004; Interviewee 17, 2005). For example, Mwanza (2005) argued that 'it should be linked . . . one hopes that the Initiative comes in to clarify gaps and do an analysis of what is going on in the sector, then after the gaps have been identified we need the Facility to fill up those gaps.' Similarly, Monique Le Génissel (2004) of the French Department of Natural Resources and the Environment, and former leader of the IWRM component in EUWI, held that 'it is very important to make a link between the strategy, which is the European Initiative, and the tool, which is the Facility.' However, while 'The ACP-EU Water Facility provides a useful parallel and complementary mechanism to the EU Water Initiative' there were no formal linkages between the two (EuropeAid 2006). As a Commission official (Interviewee 11, 2004) explained:

> The European Commission is . . . responsible and liable for the disbursement and use of the EDF. The ACP-EU Water Facility is, in effect, a project of the EDF and it is managed by . . . the EuropeAid Office . . . the EU Water Initiative . . . [can] in no way . . . have any kind of managerial or other input into an instrument for which the Commission is legally responsible.

Nevertheless, as this interviewee (11, 2004) pointed out, 'the Water Facility won't work without the Water Initiative, and the Water Initiative is unlikely to work without something like the Water Facility.' Within Africa 'the Working Group is looking at sector strategies, and sector strategies become priorities for programmes and projects, and out of that come opportunities which then need to be funded' (Interviewee

11, 2004). Thus, while the presence of EUWI in an African country did not guarantee that funds would be made available through the Water Facility, there was a degree of linkage in the fact that the Facility should be in line with the priorities and objectives of EUWI.

However, participation in EUWI processes did not bring with it a promise of financial assistance from the Water Facility: 'The WF allocates its resources on a fully competitive basis . . . Countries therefore cannot be given any guarantees of support from the WF, even if they have been subject to an EUWI-initiated dialogue process' (EUWI 2005d, 4). But there was considerable misunderstanding of this point. For example, in interviews with Zambian stakeholders, they often talked of the Initiative and the Facility interchangeably, essentially viewing them as the same thing (Chisanga 2005; Interviewee 19, 2005; Kadimba-Mwanamwambwa 2005). Furthermore, many African actors did perceive involvement in EUWI as a means to obtain funds. Annette van Edig (2004) reported in November 2004 that she believed 'the Africans are expecting some sort of project pipeline . . . which means money, and that is not what the EU Water Initiative is going to do because there are no resources.' Another EU Member State representative (Interviewee 12, 2005) observed that 'the way a lot of people from Africa look at it . . . the Working Groups of the Water Initiative [are] . . . a way to get access, or preferential access to the Water Facility.' Osward Chanda (2005), Director of the National Water Supply and Sanitation Council (NWASCO), the regulator of water supply and sanitation providers in Zambia, (who shared offices with the German development agency GTZ which coordinated the EUWI process in Zambia) said of the Facility and the Initiative that 'I can't separate the two. I would say there is commitment to the dialogue with a view to getting funding.' A MEWD official (Interviewee 19, 2005) reported that his ministry put forward:

> two proposals to the Water Facility . . . getting support for these proposals is one of the ways we expect to gain on the ground from this initiative [the EUWI] . . . The EUWI and the Water Facility come together in cohesion . . . Hopefully Zambia's involvement in the dialogue will give us more of a chance of getting funds . . . Before there was an opportunity to apply for funds then it was just another forum without making any major physical contribution, but now with the call for proposals this could actually do

> something . . . but unless funds actually flow and things happen on the ground then there may be a feeling of demoralization.

Moreover, a donor representative in Zambia (Interviewee 20, 2005) warned 'that until people see that money has arrived, it [EUWI] will never take off.'

This confusion in Zambia between EUWI and the WF was a serious problem for EUWI, though it was not an issue confined to Zambia. The EUWI Secretariat (2006, 2) reported that the fact that 'the EUWI is not, as and of itself, a provider of resources . . . has been, and still is, difficult to communicate to EU partners and stakeholders.' Furthermore, 'AMCOW members (and some other stakeholders as well) are confused about the difference between the Water Facility managed by the EC, and the multi-stakeholder EUWI for which the EC acts as a Secretariat' (EUWI Secretariat 2006, 2). To some extent, EUWI was itself responsible for the confusion, by declaring that the ACP-EU Water Facility was 'an instrument of the EUWI, helping to fund actions of key importance. One project is now being prepared for the WF to provide about €2,6 million over three years to strengthen AMCOW, another smaller project has been approved to facilitate the country dialogues in selected countries, a third project to cover costs of the EUWI Africa Working Group is being considered' (EUWI Secretariat 2006, 3). This statement exemplified the mixed messages reaching Africans about how the ACP-EU Water Facility and EUWI worked and related to each other.

WaterAid and Tearfund (2005, 3) said that the confusion had damaged the work done by EUWI. They noted that 'most partner countries . . . assumed that the Initiative did include resources,' and that since the Water Facility came on the scene, there had been:

> widespread confusion about the way the EUWI and the EUWF [EU Water Facility] relate to each other . . . The EUWF has actually had a negative effect by distracting hard-pressed officials. They have focused on submitting proposals – the vast majority of which will be unsuccessful since the EUWF was 15 times over-subscribed – rather than, for example, on lobbying their own Governments to prioritise the water and sanitation sector more in the allocation of debt relief monies or aid increases. This is exactly what has happened in the EUWI Ghana dialogue.

A Member State representative (Interviewee 13, 2005) confirmed this judgement, explaining that:

> the country level dialogue workshop in Ghana . . . was at the time when the call for proposals for the Water Facility came out and totally took over. Everything became a discussion about how we access that money rather than how we improve working together to improve efficiency in what we are already doing together . . . I have no doubt that the Facility is taking the front seat.

The September 2005 final draft of the *Strategy for Development of the EUWI* (EUWI 2005e, 5) stated that 'It will be possible to make progress only when it is understood that the EUWI is a mechanism for harmonisation, coordination and alignment of development resources but not a provider of such resources in its own right.' Indeed, some participants believed that the Water Facility overshadowed EUWI, in that commitment to the ACP-EU Water Facility outweighed commitment to the EUWI (Interviewee 17, 2005). However, the opposite view was expressed by Werchota (2005) with regard to Zambia: 'if the EU Water Facility starts projects here then I think the EU Water Initiative will also get more prominence and it will be easier for us to push forward.'

Zambia put forward seven proposals in response to the first call for proposals from the Water Facility, when the first tranche of €250 million Euros was to be distributed. Four proposals were approved and one went on a reserve list (which meant it was technically approved but there were not enough resources to cover it). Thus in total, Zambia was to receive €10,249,020.20 from the first tranche of the ACP-EU Water Facility. It remains to be seen how great an impact this funding will have in its own right but, as of January 2008, the fact that Zambia had obtained these finances failed to reinvigorate the EUWI process in-country. Indeed, it may be that the EUWI is now seen as insignificant in comparison with the Water Facility.

Non-Governmental Stakeholder Involvement

Another issue raised by the Zambian country dialogue is that of non-governmental stakeholder involvement. The *Summary and Lessons Learned* document produced on the Zambia country dialogue stated that the process aimed to give 'all relevant stakeholders a voice to articulate their perspectives and needs for improvement' (A-EUWI 2005, 1). An EUWI Steering Group Discussion Paper maintained that 'The EU is committed to ensuring that partner country governments engage civil society stakeholders in these dialogues, be they international or national NGOs, local community organizations, religious organizations, professional associations, trade unions, academic representatives and others with an interest and role in water resources management' (EUWI 2005d, 4–5). Some of the above groupings were engaged in the country dialogue process – international NGOs and one professional association. For example, WaterAid and Oxfam representatives attended the May 2004 workshop. Care International and UNICEF were also involved, but to a lesser degree – a representative from each organization met with the consultants when they were compiling their reports. A representative from the Water and Sanitation Association of Zambia, was also interviewed as part of the consultancy work, but again did not attend the final workshop.

Sievers (2005) affirmed, as an EU Member State representative in Zambia, that 'we are really trying not to exclude anybody' in the country dialogue process. A WSP and World Bank representative (Interviewee 22, 2005) agreed, stating that 'I think they did try to get all the key players there [to the national workshop], whether they actually succeeded is another matter.' Similarly, Osward Chanda (2005), Director of NWASCO, stated that 'Everyone who needed to be invited was invited and the only people not involved were those that chose to stay away.' Chisanga (2005) concluded that the process was inclusive of Zambian stakeholders because 'if you look at the water sector . . . the two ministries have been represented and the consultation did not only tackle government but also civil society and the private sector.'

However, an international NGO representative (Interviewee 17, 2005) held that while 'International organizations like ourselves . . . have been involved . . . local civil society have not been engaged at all.' A DCI-Zambia representative (Interviewee 16, 2005) claimed that engagement at the community and district levels was poor, in that local civil society 'are the key constituency that needs to be consulted but they're not the first group that are being consulted. They have a key role but I don't think they're getting a chance to play that key role.' However, this interviewee believed that this was 'not the fault of the initiative, that's the situation in-country,' because in Zambia civil society organizations which focus on water and sanitation are weak and few. As Chanda (2005) pointed out, 'Zambian based NGOs are fairly small and not known much yet.' Similarly, Sievers (2005), while accepting that local NGOs and CBOs (community-based organization) should be involved, explained that he did not know of any working on water: 'There are some, but to find out whether they are serious is difficult.' Cledwin Mulambo (2005), a Senior Water and Sanitation Engineer in the Zambian Department of Infrastructure and Support Services agreed, stating that, 'when I visited Uganda I found that there is a strong local NGO force involved in the water sector . . . [but] our NGOs are not moving in water.' Paul Kapotwe (2005) of Oxfam suggested that while local water-focused NGOs and CBOs were 'hardly visible,' 'what is more prevalent is more general CBOs that are doing a wider range of activities'. But as Chanda (2005) then asks, 'which CBOs do you approach?' because to invite everyone would be a logistical nightmare, and would result in too large a group of stakeholders to be manageable.

However, a WSP and World Bank representative (Interviewee 22, 2005) did identify some possible local groups and organizations to represent Zambian civil society in the water sector: 'You do have the Zambian Consumer Association . . . There are some districts which are very active as well, I think Eastern Province has very active communities, Northern Province has got very, very active Resident Development Associations. So there are some groups, whether they are actually being listened to is another matter.' Werchota (2005) also referred to Water Watch Groups, consumers, community representatives, and the Farmer's Union as 'key stakeholders', and stated that, although they were not invited to participate in the country dialogue process, 'they are consulted through different processes . . . indirectly they feed back into a system . . . [through] the donor round, the SWAP [Sector Wide Approach] etc.' Whether or not this indirect representation was sufficient is debateable, but an international NGO representative (Interviewee 17, 2005) maintained that while for EUWI, the community level might not have been the most appropriate level with which to engage (at least initially), 'there are organizations like Women for Change, and . . . others . . . [such as] the Water and Sanitation Association of Zambia [WASAZA]' at the national level, which could have been engaged with, and he pointed out that 'by excluding them you actually make them weaker.'

Kapotwe (2005) explained that Women for Change was a national NGO that was doing a lot of work 'around educating communities of their civic responsibilities and rights, as well as challenging government along the way, in ensuring that services are delivered,' so theoretically their involvement in EUWI would have been highly constructive. It is not clear whether they were invited to participate in EUWI processes, but it is likely they were not. WASAZA, on the other hand, is a professional association with a varied membership, with members from government, NGOs, and the private sector, 'including private sector consultants, [and] pump manufacturers' (Sievers 2005). WASAZA was, in fact, in its role as the National Committee of the International Water Association, engaged in the country dialogue process in Zambia, though only to a minimal extent. Chisanga (2005) of WASAZA argued that 'if you look at the objectives of WASAZA, one is to present to the government the needs of the people in the water sector,' so they could be said to have represented civil society, and thereby be entitled to play a much greater role in the country dialogue.

Moffat Mwanza (2005), the Zambian EUWI consultant, claimed that, although 'we didn't call any community-based organizations to the meeting' in Lusaka in May 2004, 'the [international] NGOs [such as WaterAid and Oxfam] that were present were representative of Zambian society.' Indeed, Kapotwe (2005) stated that as far as Oxfam goes, 'Most of what we are doing is people centred in that we are trying to promote people to have a voice, people to be in the driving seat, people to actually indicate their needs and push them through.' However, the *Zambia Sector Status Report*, compiled as part of the EUWI dialogue process, claimed that the international NGOs that were active in the water sector acted 'mostly as service providers rather than as civil society advocates' (EUWI 2004b, iii). Chanda (2005) remarked that 'more and more the NGOs like CARE . . . [and] WaterAid are becoming more like consultants . . . [or] contractors, implementing things on the ground . . . they are [not] representing Zambian civil society.' Indeed, the *WaterAid Zambia Country Strategy 2001–2005* (WaterAid 2001) admitted that 'Like most sector agencies in Zambia, WaterAid's present approach to field work constrains the extent to which communities, and in particular the women and the poorer and disadvantaged sections of these communities, control and influence the process. As a result interventions run the risk of being inappropriate and unsuitable.' Thus, while WaterAid wanted to address this shortcoming, and although such NGOs had links with communities, and could to some degree reflect their needs, they were not mandated to do so and, on the whole, did not become vehicles for advocacy.

In terms of local private sector involvement it was a similar story. The *Zambia Sector Status Report* stated that 'The private sector is strengthening as the demand for its services . . . increases' (EUWI 2004b, iii), but that it 'is mainly represented by consultants and contractors' as opposed to operators and small-scale vendors (EUWI 2004b, 19). A representative from an organization based within the University of Zambia and mainly concerned with undertaking consultancy work, the Technical Development Advisory Unit, did attend the workshop, but indigenous private sector involvement in EUWI in Zambia was negligible. However, this was not because they were deliberately excluded, but because, like indigenous NGOs, private sector operations in the water sector are weak and limited in number. Furthermore, the private sector operators who did exist considered that they had better things to do with their time than engage with such an apparently high-level political initiative. As Mwanza (2005), one of the consultants employed at the beginning of the country dialogue process, reported:

> we invited . . . the drillers [to the national workshop] but they did not come . . . [but] we have very few drillers, and those that are selling spare parts, for example, are few . . . we thought that was a gap that the initiative could try to fill – [local] private sector involvement in water supply and sanitation . . . but how do you do that?

The international private sector was also notably absent from the Zambian country dialogue process.

Donor Harmonization

Finally, there is the issue of donor harmonization. As we noted, four action areas were identified by EUWI for phase two of the country dialogue in Zambia: to assist in improving the prioritization of the sector; to help overcome conflict within the sector; to improve value for money within the sector; and to develop the roles of the local private sector and civil society. However, to some participants, the EUWI process in Zambia was seen as being more about EU donor harmonization. For them, as Sievers (2005) suggested, 'The initiative was basically designed to harmonize EU partners' bilateral efforts.' In the draft minutes of the August 2005 meeting of the EUWI MSF (EUWI 2005h) it was recorded that:

> The **EUWI mission** is to mobilise all available EU resources, human and financial, in a coordinated fashion in support of achieving the water-related MDGs in partner countries . . . The five [overarching] EUWI objectives [listed at the beginning of this chapter] are achieved as a result of improved harmonisation of EU efforts at regional and country levels and an increased focus of EU water-related cooperation on identified needs . . . [which are] identified as a result of policy dialogues in the regions affected . . . All actions by EU MS [Member States] to support achievement of the water-related MDGs are aligned in support of national priorities and carried out in a fully harmonised manner, with the EU speaking with one voice in all its partner countries.

In Zambia, as throughout much of sub-Saharan Africa, there is a general consensus that donor harmonization and aid alignment is a necessary and worthwhile exercise in order to avoid duplication, and to ensure that the indigenous government retains control and leadership over the donoring process. A WSP and World Bank representative (Interviewee 22, 2005) stated that 'One of the things Zambia doesn't lack is financing from donors, in a sense there is quite a lot of money that donors are putting in the sector already, but it's a question of getting that money to work effectively, and a large chunk of that depends on the donors working better together.' But while EU donor harmonization is clearly an important aim, it is not clear that a multi-stakeholder partnership is required to effect it. Furthermore, any joined-up strategy for co-ordination and harmonization in the water sector in Zambia has to include donors who are outside the EU, such as the World Bank, the WSP, the Canadian International Development Agency, and the Japanese International Cooperation Agency (JICA), because they are all involved in the sector. Hence Sievers' (2005) remark that, 'to have a really useful dialogue it has to be wider than Africa-EU so to speak.' This was a point reiterated in a document produced by the EUWI Secretariat (2006, 9) which affirmed that 'The dialogues are not an exclusive preserve of the EU and will need to involve all donors active in water and sanitation in a particular country.'

Lack of involvement of donors was illustrated by Festus Lubinga (2005), a Programme Officer for JICA, who considered JICA to be a partner in the EUWI, yet when interviewed in June 2005, stated that he did not know the purpose of EUWI and could not remember having attended the EUWI national workshop in May 2004 (although he was listed as one of the participants). However, this was not solely the fault of EUWI, as Lubinga himself conceded: 'we concentrate so much on our own activities around here. As long as other stakeholders' activities have no effect on ours we rarely get too interested.' Lubinga (2005) promised that although he had little time to spend on working on EUWI because of his other responsibilities around health issues, a new member of JICA staff would concentrate full-time on water, so that 'we will be more involved in supporting and working with the EU.'

A Harmonization in Practice (HIP) initiative was begun in April 2003 in Zambia where a HIP Framework for Action was developed which 'most of the donors in Zambia have now signed up to,' including JICA (Sievers 2005). One exception was USAID which, as one donor observed (Interviewee 18, 2006), had 'come to the table for all the harmonization meetings but . . . say they can't sign anything. They take copies of all the MoUs [Memorandum of Understanding] and they do try to align themselves [but] . . . constraints [are] placed on them from back home.' Sievers (2005), however, argued that USAID 'like to work alone and very often the way they work is very, very different . . . they spend so much money doing their own thing' that they could undermine the harmonization process. Sievers (2005) explained that the HIP process meant:

> that we coordinate our efforts, we coordinate all our programmes so that basically we can write a cheque to the Minister of Finance on budget support so that everyone's funds go in together . . . even when we have individual projects we try to harmonize whatever we do so that there aren't all these individual bilateral arrangements.

Although for donors like JICA and USAID, it was unlikely that they would be able to sign up to the idea of pooling their funding with other actors, JICA were at least open to working within a harmonized programme of activities, and even USAID in-country staff seemed willing to try and align themselves to some degree to the harmonization moves.

Following the HIP Framework, the Wider Harmonization in Practice MoU was signed in April 2004, which:

> focuses on aid effectiveness and includes an annex with specific actions and associated deadlines. These include increased use of direct budget support, establishment of more SWAPs [Sector Wide Approaches], increased reliance on government systems for procurement, fund management and auditing . . . and preparation of a Joint Assistance Strategy for Zambia (JASZ) with an improved division of labor (Aid Harmonization and Alignment 2006).

The JASZ was formulated as an instrument of aid harmonization, to be a 'one country strategy for all donors which will be a reflection of the [fifth] NDP [National Development Plan],' which was developed by the Zambian government to address the objectives and strategies for the MDGs over a five-year period (Interviewee 18, 2005). A donor agency representative (Interviewee 18, 2005) explained that within the JASZ:

> all donors are being asked to only lead in a maximum of three areas . . . the prioritization process will be done by government, donors will then have to choose among the government's priorities, also bearing in mind the financial importance the government gives each of the sectors . . . if the government wants to talk about water and sanitation, for example, it goes to one office rather than having lots of meetings and that office's job is to make sure that everybody else knows what's going on.

However, what this demonstrates, as an international NGO representative in Zambia (Interviewee 17, 2005) noted, is that with regard to harmonization and co-ordination, while 'There has been progress at a slow pace . . . the EU Water Initiative [hasn't] necessarily had anything to do with it . . . [it] was starting to happen before this initiative came into the picture.' Therefore, even if donor harmonization and improved co-ordination of the sector were originally EUWI objectives in Zambia, the processes that were already at work within the country made these objectives somewhat redundant. As Sievers (2005) commented, 'we don't really know what it [EUWI] is doing for us that we're not already doing . . . there is really no dialogue between Zambia and the EU [but] there is a lot of dialogue in other fora . . . [so] it has maybe been overtaken by events here.' But Werchota (2005) was alert to this danger, acknowledging that:

> we have to see how the EU Water Initiative fits into this process and make sure it isn't another programme which is coming on top of already existing programmes . . . the EU says we need a national programme and an activity plan, and all the other donors say the same. So we have to see how we get to this sector wide approach . . . and then see how we can use the comparative advantages of the different donors so that we optimize our input. It is not 100 per cent clear yet who will do what and that is the position the EU Water Initiative is in.

We have discussed at length the *processes* of EUWI. It is time now to examine what EUWI has achieved in terms of *outcomes* in its work on water and sanitation in Africa.

Evaluation of Outcomes

While the processes of partnership and the concrete achievements of EUWI are equally important to our understanding of this multi-stakeholder partnership as a governance mechanism, this chapter has so far focused on discussion of processes rather than results. The preponderant weight given to processes is because the tangible achievements of EUWI in meeting its own objectives, and realising the water and sanitation MDGs, so far have been few. Nevertheless, EUWI's outcomes is a topic that we must explore, and in doing so, the discussion in this section will be divided into two parts: first, the outcomes of EUWI in our case study, Zambia; and second, the outcomes of EUWI as a whole, with respect to its Africa-focused work.

Outcomes in Zambia

The December 2005 report, *An Empty Glass*, by WaterAid and Tearfund, two of the NGO partners in the EUWI, was highly critical of the EUWI's rate of progress and success. With respect to Zambia, they stated that the 'EUWI Africa Working Group has criticized the initial work as too consultant-led and lacking country-buy-in' (WaterAid & Tearfund 2005, 2). WaterAid (2001, 7) claimed that 'The water and sanitation sector in Zambia suffers from a distinct lack of coordination and collaboration between donors/external support agencies, and between these agencies and government institutions.' EUWI had not led to better, more efficient, and more coordinated use of EU aid (which it was originally established to do), let alone harmonised funding more effectively within the sector as a whole. In fact 'No changes in EU donor activity' within the country had occurred as a result of the EUWI (WaterAid & Tearfund 2005, 2). This was despite the fact that, as a Commission official and EUWI Secretariat member (Interviewee 11, 2004) declared, 'There are probably about 12 Member States who have significant development programmes, and who have bilateral programmes, and those, along with the Commission, are one of the main vehicles the Water Initiative needs to be able to influence.'

Although this interviewee (11, 2004) acknowledged that harmonization was 'quite a long-term [and] difficult objective,' it must also be recognized that within Zambia, other in-country processes working towards the harmonization of the sector had overtaken EUWI. In terms of outcomes, therefore, the EUWI country dialogue process faltered within Zambia and did little towards achieving its own country-specific objectives. WaterAid (2001, 5) observed that 'There is a general lack of understanding about, and appreciation of, the significance of water, sanitation and hygiene at almost all levels of government' in Zambia, but there is no evidence to suggest that EUWI has done anything to change this. EUWI did not give water and sanitation a greater profile in Zambia or strengthen political will towards achieving water supply and sanitation targets. While there are promising signs that conflict within the water sector between the MLGH and the MEWD is abating, EUWI does not seem to have played any significant role in reducing that tension. Moreover, the role of the local private sector and civil society within the WSS sector has not been enhanced by EUWI: indeed, it seemed to ignore some indigenous stakeholders. Even Werchota (2005) (of GTZ who coordinated the Zambian process on behalf of the EU) admitted that EUWI had made little impact: 'Does the water initiative have an impact on the Zambian structures? I would say at the moment, not really.'

The only positive outcome that could be said to have come about because of the Zambian country dialogue process is improved feedback: 'the findings of the [Zambia Sector Status] report are now being fed back into the process of the SWAP, and also into the donor round . . . So the findings are impacting on the discussions going on today' within the country, even if these discussions did not take place under the EUWI banner (Werchota 2005). Furthermore, lessons were learnt from the Zambian experience and this should help EUWI to reshape its future efforts elsewhere as well as in Zambia itself.

Nonetheless, even if the initiative had achieved more outcomes in Zambia, monitoring tools for EUWI were very weak, and formally non-existent throughout the initiative, an issue which we will return to in the next section. Werchota (2005) insisted that the solution to this problem lay with AMCOW and not with the European Member States: 'what I recommend is that AMCOW asks the countries to report . . . but they can only report if AMCOW tells them what they should report on.' However, monitoring is not just AMCOW's problem but a shared problem that required a joined-up solution by the TTP system as a whole.

Outcomes in General

Beyond Zambia, what have been the outcomes of EUWI? One possible outcome was the creation of the ACP-EU Water Facility. Although the Water Facility had no formal link to EUWI, it has been argued that its creation was 'a direct outcome of the EUWI' and would not have happened otherwise (EUWI Secretariat 2006, 1). Jacques Labre of Suez (2004) claimed that another outcome of EUWI was that 'the quality of the dialogue between European NGOs and private operators is greatly improved. We are now in the situation where we can write letters together to the Commission for a joint meeting with them, which would not have been possible two years ago.' Another private water operator representative (Interviewee 8, 2004) testified that:

> a big achievement of the Water Initiative . . . is that the different actors started to talk to each other. It was amazing to see the NGOs start to talk to each other. There are big differences

> between WWF [World Wide Fund for Nature] and Tearfund for example, and they started to talk to each other, send messages, and have frequent meetings, and represent each other. The [private] operators do this as well, but also the NGOs and the private operators are talking to each other, exchanging views and working together and co-operating with regards to the initiative, which in the water sector is something very amazing.

Confirmation of this improved dialogue came from an NGO representative (Interviewee 10 (2004).

A third outcome was a much clearer focus on the purpose of the EUWI country dialogue approach. Mwanza (2005) suggested that 'The likelihood of it [the EUWI] being successful, not only in this country [Zambia], but in the continent as a whole will be small if it does not clarify what it will do,' and this issue was addressed. As we learned earlier, when the Netherlands took over the coordination of the Africa WSS Working Group from Denmark in November 2004, the country dialogue approach was redefined, taking into account lessons from the relative failures of the previously begun processes in Zambia and Ghana. As a European Commission official and EUWI Secretariat member (Interviewee 11, 2004) reported, at the fifth session of AMCOW in November 2004, the African ministers endorsed proposals to change the country dialogues away from the original format of a fairly loose process whereby problems and gaps within the water sector were identified, towards a much more specific focus on developing 'sector strategies matched to the MDGs . . . [where the country dialogue] is a long-term process with an annual or appropriate review revisit, and . . . alongside that long-term sector plan, [there is] some kind of broad financial strategy which allows that sector plan to be achieved within whatever constraints are set at the national level by the national budget.' To achieve this, the EUWI Finance Working Group agreed to cooperate with the Africa Working Group 'in order to integrate at a country level some of the activities of both working groups' (Batz & van Koppen 2006). Thus the new strategy was 'trying to make something loosely defined as a country dialogue into something which has a much clearer outcome and a much clearer . . . process' (Interviewee 11, 2004).

The *Country Level Dialogues Key Documents* (EUWI Working Group on Water Supply and Sanitation in Africa 2005, 2–3) brochure outlined the objectives for the reformed dialogue process as to:

- Contribute to the achievement of the water, sanitation and hygiene MDGs in each of the countries involved;
- Improve coordination of work; strategic planning and the prioritising of efforts in the water, sanitation and hygiene sector by linking with existing initiatives to reach the MDGs;
- Rationalise PRSPs and related strategies and plans to ensure that investment in the sector is better targeted on the poor and most vulnerable and able to attract more financing to reach the water-related MDGs;
- Bring all water, sanitation and hygiene stakeholders together to identify policy, institutional, and financial bottlenecks that impede achievement of the water, sanitation and hygiene MDGs, which are then to be reflected in a national water, sanitation and hygiene Roadmap to 2015;
- Support the initiation of a national water, sanitation and hygiene Multi-Stakeholder Forum;
- Use the Roadmap as a basis to improve the efficiency and targeting of current financing, mobilise funding and achieve concerted action towards achieving the water, sanitation and hygiene related MDGs;
- Support and monitor the MSF in the implementation of the Roadmap, identify and troubleshoot problems arising, encourage the systematisation of lesson learning;
- Disseminate results and best practices to countries in the region through AMCOW;
- Continue on an annual basis as an integral part of water sector budgetary planning until 2015, the target year for the MDGs, or until the water-related MDGs have been achieved, whichever comes first.

The brochure explained that the revised country dialogue was 'an attempt to fundamentally change the relationship between donor, government and other stakeholders and to form a basis for an ongoing, effective and fruitful national level dialogue as part of a flexible and learning focused approach to achieving the MDGs' (EUWI Working Group on Water

Supply and Sanitation in Africa 2005, 3). This document was much clearer than before on the detail of how the initiative should operate at the country level, stating that an MSF would be established within each country, facilitated by the national government with the support of the respective EU Member State. This would be 'the engine of the Country Dialogue process . . . [and] will bring together all the principal stakeholders in the sector: government, donors; NGOs, private sector and civil society . . . The MSF will meet at least annually but will also establish working groups to deal on a functional level with specific issues' (EUWI Working Group on Water Supply and Sanitation in Africa 2005, 3). The brochure listed seven phases of the new dialogue process, describing the objectives and activities of each phase in detail, as well as which actors were responsible for its completion: Phase 1 – Preliminary AMCOW-EUWI Mission; Phase 2 – Drafting of an inventory as a basis for the Multi-Stakeholder Forum discussion; Phase 3 – Establishment of a Multi-Stakeholder Forum; Phase 4 – National workshop to discuss and formulate a National Roadmap; Phase 5 – Endorsement of the National Roadmap to 2015; Phase 6 – Roundtable on Roadmap Financing; and Phase 7 – Annual Progress Review (EUWI Working Group on Water Supply and Sanitation in Africa 2005, 4). As a result, the country dialogue processes in future would be much more structured and focused, with clearer lines of accountability than before.

However, progress was painfully slow. Eleven country dialogues were planned (the Democratic Republic of Congo was added to the original list of ten pilot countries), and 'the water policy dialogues will gradually be extended to include an increasing number of countries identified jointly by the EU and its regional partner organisations as off-track with regard to achievements of the MDGs' (EUWI 2005e, 3). But by the end of 2005, country dialogues had started in only five countries and were 'set to commence in three others . . . as well as the DRC, (not originally a pilot country), while there has been no progress in two pilot countries' at all (EUWI Secretariat 2006, 4). Since then, progress has still been very slow, as the *European Union Water Initiative Annual Report 2007 – Draft* (EUWI 2007) highlights. Here it is reported that in 2006 the initiative 'focused on reinforcing a good dialogue between the African Ministerial Council on Water Technical Advisory Committee (AMCOW-TAC) and the EU . . . [and] several meetings were held . . . to define an activities work plan 2006–2008' (EUWI 2007). However, 'The discussions on the new strategy and the related work plan for 2006–2008 had to be concluded before financing for the activities could be secured' which had 'resulted in very few concrete activities of the Africa Working Group in 2006' (EUWI 2007). As for the country dialogues, the report stated that, while there had been 'mixed success' in developing these overall, a successful process had been launched in Ethiopia in 2006 and dialogues in Congo Brazzaville and the Central African Republic were scheduled to begin in the summer of 2007 (EUWI 2007). Furthermore, the report noted that 'the Africa Working Group has determined that there should be a more selective approach to Country Dialogues' in future, and they will therefore 'launch an evaluation of those Dialogues that have been initiated through earlier activities' (EUWI 2007).

In 2005, WaterAid and Tearfund (2005, 2) attributed the limited rate of progress to a lack of commitment by EU Member States:

> Three years on, this raises questions about the vigour with which these dialogues have been pursued. The main delay has been the time taken for a member state to volunteer as lead donor on behalf of the EU in the country dialogue . . . [which demonstrates the fact that] The vast majority of member states are not sufficiently committed [to EUWI].

Lack of commitment on the part of the Member States was also recognized at the fourth EUWI Steering Group meeting in October 2005 where Germany criticized the new draft strategy paper for the initiative, claiming that 'There is a problem of political commitment from the EU MS, but the strategy paper should not ask for more commitment, creating the impression that the water sector is unable to attract the attention of decision makers' (EUWI 2005a).

As for outcomes in terms of actual delivery of water and sanitation services, an international NGO representative (Interviewee 10, 2004) contended in April 2004 that no one 'in the EU Water Initiative could point the finger at any single individual that has had direct benefits from the initiative that they wouldn't have got before. If that situation pertains for another year to 18 months then I don't think it will go anywhere and it won't make a difference really at all.' Similarly, WaterAid and Tearfund (2005, 1–2) claimed that: 'Not a single extra person has received safe water or sanitation through the Initiative . . . the EUWI has not succeeded in accelerating delivery of the water and sanitation MDG targets.' Moreover, there were no mechanisms in place to calculate whether or not such concrete outcomes had occurred. As Interviewee 10, 2004) contended in April 2004, it would have been 'very difficult to judge any success because the . . . reporting is awful.' Indeed this interviewee (10, 2004) claimed that 'there are quite a few Member States that are . . . utterly resistant to monitoring what they're doing on the EU Water Initiative. There isn't a proper strategy and there aren't clear objectives but you need some baselines to measure change, to be able to know where you should be going to.'

With regard to monitoring, at the eighth MSF in June 2003 it was recognized that 'there is a need to agree what the EUWI is meant to be achieving in terms of objectives and have means to monitor its progress . . . Without a clear idea of the expected outcomes of the EUWI it will be impossible to monitor [the] progress and impact of the Initiative' (EUWI 2003c, 3). Therefore, it was agreed that a sub-group of the MSF would be formed 'to prepare a paper on monitoring objectives and methods' (EUWI 2003c, 3). At the tenth MSF in March 2004 'The Commission presented the position paper on monitoring and reporting' (EUWI 2004a, 1), and it was decided that a Working Group was needed to 'Outline an effective M/R [Monitoring/Reporting] system for the EUWI' (EUWI 2004a, 2). Interviewee 10 (2004) claimed that, in effect, the Commission 'forced' the Member States to have a Working Group on monitoring. However, at the end of 2005, WaterAid and Tearfund (2005, 2) reported that accountability within EUWI was still poor:

> States get away with not participating fully in the Initiative. The EUWI is mentioned in the EC Annual Report but there are no indicators for monitoring its progress. The [Monitoring and Reporting] working group . . . is chaired by the Italians whose commitment is unclear. Their initial proposals were very weak. The EUWI's progress, or lack of it, is therefore kept well away from the public scrutiny.

This led these EUWI participants to state that 'The EUWI's major problem is the absence of accountability which is in turn driven by a lack of transparency and limited participation of Southern governments and civil society' (WaterAid and Tearfund 2005, 5).

A serious attempt to get to grips with this problem was set out in September 2005, as the *Final Draft Strategy for the Development of the EUWI* (EUWI 2005e, 8) noted:

> A comprehensive monitoring and reporting system is being developed by the Monitoring and Reporting Working Group, which will allow assessment of the impact of the EUWI as an outcome of the EU MS' [Member States'] involvement. The system will have four types of quantitative indicators (simple indicators, aggregated indicators, indicators of advancement and coverage ratios) and a set of qualitative/participatory indicators based on questions to EUWI partners and stakeholders in order to derive qualitative scores . . . Financial assessments of the efficiency of the EUWI and its results in partner countries in terms of impact on provision of water and sanitation for the poor will also be available . . . Implementation of the system will involve all the main EUWI actors and provide the assessments needed for regular review of the EUWI strategy.

This proposed monitoring system even included an evaluation of the extent of stakeholder participation. As we have learnt, non-governmental stakeholder participation in EUWI has not always been inclusive or equitable, despite the assertions that 'the central feature of stakeholder involvement in the EUWI should not be in doubt' (EUWI 2005d, 5), and that 'Stakeholder involvement will be one of the areas where the eventual success or failure of the EUWI will be assessed' (EUWI 2005d, 5). Several indicators were to be used to monitor this aspect of the initiative, with an 'emphasis on . . . relative change rather than absolutes' to take into account different contexts (EUWI 2005d, 5). Moreover, each Working Group was asked to prepare a Work Programme for 2006, which the WG chair was expected to complete, preferably with the involvement of 'partners and stakeholders' (EUWI 2005a). To be included in the Work Programme was an explanation of proposed activities for the year, the expected output of each of these activities, how these activities link to WG objectives, as well as a timeframe and cost. At the Steering Group meeting in October 2005, 'There was a presentation of the monitoring system by . . . the chair of the WG on Monitoring and Reporting' which was to be applied by the chairs of WGs (EUWI

2005a). In April 2006, a draft version of the Africa WG Work Plan became available on the EUWI Communication and Information System, though only for members with access to this internal system (EUWI Africa Working Group 2006a). By the summer of 2006, 'a methodology for the Monitoring of the EUWI' was finalized and a 'first Monitoring Report was presented to the Multi-Stakeholder Forum in Stockholm' in August 2006 (EUWI 2007). It remains to be seen how effective this methodology will be and how well it will be embraced by the other Working Groups within the EUWI, but at least there is now an established framework for monitoring.

These attempts to establish a monitoring system were precipitated by increasing doubts about EUWI's effectiveness. In June 2005, at a meeting in Addis Ababa, the Executive Committee (EXCO) of AMCOW 'called for a mid-term review of the EU-Africa Strategic Partnership on Water and Sanitation in order to assess its progress and establish ways of accelerating the implementation of the EU Water Initiative' (Batz & van Koppen 2006). This was 'the result of their perception that the progress over the last three years was slow and the EUWI was not meeting the expectations of major stakeholders' (Mushauri 2006, 5). Moreover, EUWI was not perceived to be 'delivering new investments' (EUWI Secretariat 2006, 2). This review was to take the form of a facilitated workshop in Uganda on the 13–14 February 2006, looking at both the EU and the AMCOW perspectives. Here it was noted that 'AMCOW considers that the EUWI was not established as a result of common agreement on the needs of the African water sector and may not be the most appropriate way to address these needs' (Ntale & Swann cited in Mushauri 2006, 19). Furthermore, Reginald Tekateka (cited in Mushauri 2006, 10) of the Department for Water Affairs and Forestry (DWAF), South Africa claimed that:

> There are serious considerations currently in Africa regarding the way forward on the EUWI. The considerations are around reviewing the partnership, renegotiating the partnership or withdrawing from the partnership altogether. Most serious concerns from the African perspective are around:

- Lack of ownership as the initiative seems obviously Euro-driven
- Lack of sufficient resources to run the dialogue process itself and

- Unclear roles and responsibilities and accompanying processes and procedures for getting things done.

It was therefore agreed that:

> The Country Dialogue process is a valuable tool, but,
> - they must be demand led.
> - they must involve all relevant actors (and not be the exclusive preserve of the EU).
> - the EU will seek funding for those of the 10 pilot countries that have not started the dialogue process. A predictable source of financing for the activities must be agreed if the dialogue process is to be continued (Mushauri 2006, 22).

And furthermore, that:

> A forum will be established between AMCOW and EU at the level of water sector experts to monitor the progress of the Partnership. The forum will meet at least annually, preferably in Africa and attached to AMCOW meetings. The forum will develop a set of indicators which will be used to monitor progress . . . .The forum will report to AMCOW-EXCO and the EU at the political level as part of the high level political dialogue between EU and Africa (Mushauri 2006, 22).

However, it was not made clear if the reviewing structure suggested here would supplant the previously planned system or be in addition to it. It was generally agreed that monitoring of the initiative was vital and should not be undertaken by European actors alone, but it remains to be seen whether the proposed monitoring system will be effective and functional, let alone whether there will be any significant outputs for it to monitor.

Conclusion

In this chapter on our first case study – EUWI – we have shown how it is a very broad partnership, with a multitude of participating organizations. Focusing on its work in sub-Saharan Africa, with special reference to Zambia, we discussed, first, its northern context – i.e. the contributions made by partners from Europe, including EU Member States, the private sector, and NGOs – and concluded that it is a northern-dominated and government-led partnership. Second, we discussed its regional or EU-Africa context – i.e. the contributions made by African governments, the indigenous private sector, and indigenous NGOs and CBOs – and concluded that although EUWI is northern-dominated, African governments are genuine partners, but neither the indigenous private sector nor indigenous NGOs and CBOs have been very visible. Third, in the Zambia case study, we found that EUWI's country dialogue process, designed to engage all stakeholders in decision-making, was bedevilled by anchorage issues (how to embed the initiative in an indigenous ministry or other local host) and harmonization problems (how to avoid duplication of donor partner's activities). Also, there was confusion over EUWI's objectives; its relation to the EU Water Facility; and the connections between its operations in Zambia and other layers of EUWI's structure. Running through all these issues is uncertainty over the role of non-governmental stakeholders, and lack of machinery for monitoring the progress of EUWI in meeting its goals – whatever they are. We now turn to our second case study – PAWS – where we will find many of the same issues illustrated.

4

PARTNERS FOR WATER AND SANITATION (PAWS)

Introduction

In the year before the WSSD, the UK Prime Minister, Tony Blair, urged private companies and NGOs to work with government on initiatives in five areas: forestry, tourism, financial services, energy, and water. As a result, PAWS was launched in December 2001 by Michael Meacher, the UK Minister of State for the Environment, at the International Freshwater Conference in Bonn. In its original conception, this initiative was to focus solely on water, but later a sanitation element to the partnership was added. PAWS featured in the WSSD Water Dome and became registered with the CSD Secretariat as part of the WSSD process.

PAWS, a not-for-profit initiative, comprises approximately 38 UK-based members: from the government side, one government agency and four UK government departments; from the private sector, almost all of the UK private sector water operators, three consultancy firms, five trade and professional associations, a law firm, and a marketing and communications company; and from civil society, a university, a trade union, and four charitable organizations. In addition, by the time of the WSSD, PAWS had established three sub-Saharan African government partners – South Africa, Nigeria and Uganda. In January 2005, a MoU was also signed with the Ethiopian Ministry of Water Resources, and in September 2006 a country manager commenced work in Ethiopia (PAWS 2006k).

The PAWS initiative's overarching aims were stated as (PAWS 2006a):

- To deliver tangible and sustainable benefits that make a recognisable and verifiable difference at the local community level.
- To emphasize strong inter-linkage between water supply and sanitation.
- To encourage best practice and develop guidelines for effective and sustainable tri-sector partnering.
- To focus attention to secondary towns and peri-urban areas in Africa.
- To demonstrate the effectiveness of multi-stakeholder engagement.
- To promote socially and environmentally sustainable services that endorse principles of good water management.
- To champion an environment that attracts investment and ensures financial sustainability.

An NGO representative (Interviewee 1, 2003) declared that essentially 'the aim is to demonstrate that by working together in partnership in the UK we can use the skills and shared experience that that produces, to work within certain locations . . . [where local actors can see what they] can draw from the partnership that is relevant' to them. Jeff Davitt (2004) of Yorkshire Water (part of the Kelda Group) claimed that the objectives of the partnership were two-fold: 'to increase provision of water and sanitation services . . . for people in Africa, but also to demonstrate the benefits of the partnership approach. So one is achieving the fundamental goal of increasing provision but . . . [alongside this the aim is] to increase learning and create something which can be replicated elsewhere.' Similarly, Nicola Clarke (2005) (who was a member of the PAWS Secretariat at the time of interviewing) contended that:

> PAWS's objectives [are] split in two, you've got grassroots objectives which are the 'lets do something' to deliver water and sanitation at the local level by building capacity, and then there are two objectives which basically say we also have a role to try and figure out whether partnerships themselves can work, and to pass on knowledge . . . the grassroots bit, obviously is what's really important, water and sanitation is key, but . . . if you don't get the

> bit about how the partnership works itself right, you're not going to achieve anything either.

On the first set of objectives (grassroots engagement), although originally, PAWS 'was going to be about delivery' of services (Mitchell 2003), this idea was later dropped, and, as many participants reported, the PAWS partnership's main focus became capacity building (Interviewee 3, 2003; Interviewee 14, 2004; Mitchell 2003; Peacock 2003; Rouse 2003). As a private sector participant (Interviewee 14, 2004) stated, 'PAWS is about . . . delivering sustained capacity building.' Paul Turner (2005), a Northumbrian Water secondee to the PAWS Secretariat, suggested that it was about 'access to the knowledge base [of the UK partners], knowledge transfer, mentoring . . . [less about] technical knowledge . . . and more [about] organizational and management issues.' An NGO representative (Interviewee 1, 2003) said: 'it's information rather than project based, it's [about] information access . . . [which is] unusual.'

Les Peacock (2003) (who, when interviewed, was head of the PAWS Secretariat) similarly stated that:

> the emphasis is on building capacity with the people in the municipalities, not going in, doing something [physical like digging a bore hole and putting in a pump, and then] going out . . . [because] all over Africa there are examples of good projects, with probably the best intentions, and they fail because they either break or people . . . steal [the fittings].

Other reasons for failure of previous approaches include: a lack of a sense of ownership over projects from the community and indigenous government; failure to meet the communities' needs; inappropriate technology use; and a lack of capacity to maintain and run water and sanitation services once a donor agency or group has left. The idea behind PAWS was that a new approach was needed, to avoid and/or address the problems encountered by water and sanitation projects in the past, through multi-stakeholder, demand-led work, which sought to build capacity. And 'While the initiative does not itself provide funding,' it aimed to strengthen 'each locality's ability to identify and access available sources through the capacity building approach.' As the PAWS *Annual Report 2002–2003* (PAWS 2004b, 5) noted, 'The emphasis of partner involvement is on capacity building to ensure the sustainability of projects and that they can be replicated. Particular areas of capacity building could include corporate or institutional, financial or technical.' The type of project work PAWS would undertake, however, was a cause of some confusion amongst members, as we will see later in this chapter.

On the second set of objectives (partnership development), Michael Rouse (2003), President of the International Water Association (a network of water professionals and charity by status) maintained that PAWS was about 'developing models' of partnership using 'case study material . . . to try and persuade aid agencies that this is the way you can make progress . . . it's very much about establishing demonstration projects . . . [to show how] you can use the skills that are available . . . [in] the UK' to help developing countries deliver water and sanitation to their people. Similarly, a Thames Water representative, Ed Mitchell (2003), contended that 'the original idea was that we would prove some sort of new approach, we would prove the value of partnership working essentially, and by proving that we would be able to take it to scale . . . [and work] in a lot more places, we'd be able to attract money.' However, Mitchell claimed in November 2003 that since it began, 'PAWS has sort of reined in its ambitions,' and did not have 'the appetite to scale-up' its work that it once did. Instead, he suggested, PAWS was seeking 'to demonstrate the value of their contribution to those municipalities' that had been already been identified for it to work with (Mitchell 2003).

The partnership development undertaken by PAWS was to be achieved through three key strategies, listed by the partnership as follows (PAWS 2006h):

1) We provide strategic support on all aspects of sustainable water supply and water resource and sanitation management. This includes governance structures, all forms of regulation and socio-economic strategies. This support is based on broad knowledge of operational models, including the development of policies and institutions for the delivery of water supply and sanitation services and the integrated management of water resources.
2) We transfer knowledge by twinning local water service organizations and municipalities with similar organisations in the UK. In addition we organise networking events, such as targeted seminars

and inward and outward working visits at the secondary town and peri-urban levels.

3) We identify and prioritise projects by assessing the needs of the country through in-country stakeholder groups. These projects can be technical – covering leakage strategy, sanitation strategy and business planning – or related to the socio-economic aspects of capacity building – customer care, stakeholder engagement and revenue collection. These are generally short-term projects that may facilitate and support longer-term action and support.

This chapter is structured like chapter three, beginning with the PAWS partnership in the northern context, focusing on the processes and modalities of its working in the UK. Then we consider the southern context, focusing on the partnership between UK actors and those in PAWS's African partner countries, examining work in three specific countries, South Africa, Uganda and Nigeria, with special attention given to South Africa. In the remaining two sections, we evaluate the outcomes of PAWS as a TTP, and conclude with a summary of the findings of the chapter.

The Northern Context

The PAWS partnership was administered and supported by a Secretariat housed in the Department for Environment, Food and Rural Affairs (DEFRA), which provided most of the funding for PAWS (with some contribution from the Department for International Development (DFID)) up until the end of March 2007, and until 2005, DEFRA chaired all meetings. However, DEFRA decided that it could no longer provide funding for the initiative at the end of the 2006/07 financial year, and the Department for International Development (DFID) took over as the lead funding body until at least 2010 (PAWS 2007). At this point, the Secretariat also moved out of DEFRA to be hosted by the Water Engineering Department at Loughborough University (PAWS 2007).

The PAWS website (PAWS 2006h) stated that the Secretariat was responsible for:

- Identification and delivery of projects in partner countries
- Resource management of UK partners

- Programme and project management
- Identification of potential partners and countries
- Partner management and communication to an agreed strategy

During the first three years of the partnership's history, there was a Forum which met twice a year, to which all members were welcome (PAWS Secretariat 2004). After the establishment of the Forum, a Steering Group (which was to report back to the Forum) was appointed, 'comprising representatives from each sector of the partnership' (PAWS 2004b, 4). In theory, the Steering Group was to meet more regularly than the Forum, and be responsible for devising and co-ordinating partnership policy and formulating strategy for the involvement of partner countries. In practice, however, the Steering Group became incorporated into the Forum (PAWS Secretariat 2004). In February 2005, Terms of Reference (ToR) for PAWS were adopted which retained the concept of a Forum, but which met only annually, and 'at which meeting it . . . formally appoint[ed] the Chair and Steering Group' (PAWS 2005a, 3), and signed-off 'the business plan for the following year' (Turner 2005). The formalization and creation of a Chair outside government was a new requirement stipulated in these ToR, and the Chair's 'function is to provide strategic leadership of the Partnership and ensure that PAWS is effectively managed' (PAWS 2005a, 2). The Steering Group met quarterly, chaired by the PAWS Chair, and was 'responsible for overseeing implementation of the business plan,' with its membership made up of 'at least one representative of the three sectors,' government, the private sector, and civil society (PAWS 2005a, 3).

In the original conception of PAWS, the UK government was to provide seed money and initial secretarial support for PAWS before it evolved with 'a life of its own' (Mitchell 2003). However, six years after the partnership was launched, the UK government continued to provide both the Secretariat (housed in DEFRA) and funding for the partners' subsistence and travel costs, although Paul Turner (2005) of the Secretariat insisted, in June 2005, that DEFRA were still hopeful that eventually 'the partners . . . [would] ultimately take it on so that it has a life of its own and with a Secretariat independent from government.' Interviews with a cross-section of PAWS members in 2003 and 2004 suggested that there was a general consensus that PAWS was largely a northern government-led initiative with most meetings taking place in civil service offices in London, chaired by the government-based Secretariat (Barker 2004; Interviewee 1, 2003; Interviewee 2, 2003; Interviewee 3, 2003; Interviewee 14, 2004; Rouse 2003; Mitchell 2003). A not-for-profit member (Interviewee 2, 2003) observed that 'it's perceived by most people sitting round the table as a DEFRA-led initiative.' Peacock (2003) admitted that although the Secretariat was there 'to support the partnership,' it was also there 'to lead the partnership and take things forward, [although] it isn't the decision-making body.' An NGO representative (Interviewee 3, 2003) said of the UK government's role that 'they're putting the money into it, they're funding it, so I think they do as much consultation as possible but ultimately they make the decisions.'

Indeed, there was a widely held impression within the partnership that the UK government was too firmly positioned in 'the driving seat' of the initiative (Mitchell 2003). John Barker (2004) of WWF contended that the process should 'be made more independent of government,' because, while 'There is no doubt that it is a multi-stakeholder process . . . it is driven by government, they are the major players,' and this meant that 'engagement from the [other] partners is [not] as full as it probably should be.' A not-for-profit member (Interviewee 2, 2003) maintained that PAWS 'could have [been] structured . . . in such a way that it would create greater equity . . . between the partners.' Members suggested that this could have been achieved through rotating the chair and/or through having secondments from civil society and the private sector to the Secretariat (Barker 2004; Interviewee 2, 2003; Interviewee 14, 2004).

The UK government's dominant role in PAWS was criticized for preventing a genuine partnership from emerging. Lack of a genuine partnership was evident from revelations that some representatives of organizations listed as official PAWS members, such as Halcrow (2003), claimed they were involved only to a very limited extent, or they were 'sleeping partners'. Since, as Interviewee 2 (2003) pointed out, 'The purpose of partnership is that everyone puts something in and everybody gets something out,' if there were "sleeping partners", then something was wrong with the way the partnership was working. Interviews with members of PAWS revealed that partners other than government lacked a sense of ownership and a perception of mutual benefits (Barker 2003; Interviewee 3, 2003; Mitchell 2003; Nowak 2003; Oudyn 2004). Mitchell observed in 2003 that:

> partnerships only work if there are benefits to all partners and I'm not sure yet that PAWS is delivering sufficient benefits to all . . . [the fact] that DEFRA have had to maintain their funding and Secretariat role . . . shows that it still needs that kind of government stimulus to keep it going, and therefore isn't achieving . . . what a really great partnership . . . should achieve.

Because the UK government assumed and maintained a dominant role within the partnership, it weakened PAWS's partnership modalities.

However, the UK's government's leading role in PAWS was defended by Christopher Causer (2005) of the law firm, Kirkpatrick & Lockhart Nicholson Graham LLP, who stated that 'There is a fair amount of public sector money that goes in to PAWS, it is funded by the tax payer so . . . quite rightly, civil servants are keen to have everything documented and properly set-out.' Similarly, Clare Twelvetrees (2005) of the Environment Agency remarked that 'they have to keep an overview of what's going on in order to justify their money being spent.' Also, an NGO representative (Interviewee 3, 2003) argued that although 'the government . . . ultimately dominates . . . [it is not] in a way that is unacceptable,' and 'the actual structure and governance has worked out well' with 'openness, transparency, and joint decision making.' Davitt (2004) claimed that 'there is more to it than just a government approach . . . all the sectors bring a different viewpoint and the fact that we've been able to achieve consensus and a way forward is a pretty strong demonstration of the success of it . . . holding the Secretariat within DEFRA keeps things moving and aligned.' Likewise, Sue Nowak (2003) pointed out that because the UK government fund and support the partnership, it is 'one thing that the partners don't have to worry about on a day-to-day basis.'

In other words, the UK government had to retain its prominent role because other stakeholders were either not inclined, or did not have the capacity, to take responsibility for these functions (Nowak 2003). Causer (2005) agreed that the Secretariat 'needs to take a lead because the rest of us have all got busy jobs . . . It's a classic civil service function and a very important one – to pull people together, circulate the papers, sum it up afterwards and send out action points.' Moreover, Rouse (2003) observed that within the partnership 'there are equal opportunities for those who want to put the effort in,' while Rachel Oudyn (2004) claimed that 'you're as integrated as you want to be.' Similarly Twelvetrees (2005) argued that, 'there are some sleeping partners [but] I think active partners share a sense of ownership.' Even a critic of PAWS's governance structure conceded that while it 'has faltered in terms of its partnership mechanisms at times . . . it's done a lot better than most' TTPs (Interviewee 1, 2003).

The quality of partnership was also affected by the way in which DEFRA conducted PAWS meetings. An NGO representative (Interviewee 1, 2003) complained that because in PAWS meetings, members always sat 'around a boardroom style table and . . . had an agenda, and had a [formal style] meeting,' participants 'never really got under the skin of each other' to talk openly about such things as perceptions, motivations and expectations. What resulted was the creation of a committee rather than a partnership, although an NGO representative (Interviewee 1, 2003) remarked that the 'real partnerships will be, if we have them, from the projects that we do.' Similarly, Causer (2005) commented that 'The stuff we do in London is fine but it's committee meetings; our real work is when we go to . . . Africa.'

The role of the private sector as a partner in PAWS raised two further issues of concern. First, representatives from the private sector were frustrated by the slow pace of PAWS bureaucracy. For instance, Mitchell (2003) stated that 'these kind of multi-stakeholder or tri-sector partnerships . . . move very slowly . . . [and] you have to bring everybody with you,' whereas, 'business is . . . quite focused, quite short-term . . . [they] go in there and do something.' Davitt (2004) said that:

> it seems to have been very slow and not got very far but I think I am getting used to the fact that when you work with all these different stakeholders and with very disparate needs from recipient countries, maybe that's normal . . . You think everything moves slowly but when you look at all the interactions and parties involved you realize it does need a lot of briefings and communications . . . and it therefore isn't going to be going at lightning speed.

The second issue was the private sector's search for business. One private sector participant (Interviewee 14, 2004) suggested that the lack of involvement in partnership activities by some members was because 'there is insufficient work for the partners to get stuck in' to. This reflected the fact that some private sector operators regarded PAWS as an opportunity to obtain work: 'At the outset some private sector actors saw this as a way of accessing contracts,' which was never the intention of PAWS (Interviewee 1, 2003). It was the consultancy firm partners, rather than the water operators, who viewed the partnership as a means to gain work. An NGO representative (Interviewee 1, 2003) explained how 'it is easier for the big companies like Thames Water . . . to see that ultimately there is this benefit for them . . . [in being] involved on their corporate responsibility side, rather than on their business side . . . but the consultancy companies that . . . [are small and] charge by the hour . . . don't have the same resources for corporate social responsibility.' The consultants 'felt they were in a difficult position because they have to be paid for their time and it's very difficult for them to give time voluntarily . . . but the [UK] government said no, the whole point of this is that it is voluntary,' (Interviewee 1, 2003). This led, if not to the complete withdrawal of the consultancy firms from the partnership, then to their disengagement from most PAWS activities. Richard Waller (2003), of the consultancy firm Mouchel Parkman, confirmed the dilemma faced by the consultants:

> there is no escaping the nub of our concern. We have something to offer in terms of knowing how to do things and what to do, experience of water projects in Africa, particularly in these countries [which PAWS is working with]. We would remain willing to make limited contributions of free time to help facilitate that process, generate projects and what have you, but I just cannot see how they can do any realistic, substantial project on the basis of piecemeal contributions from UK firms . . . it cannot work on the basis that the people you need to do these projects are going to have to do it for free. It could be set up that the initial development of the project is done by free contributions, but once you've identified a project and identified elements of work, those elements can then be contracted out.

We return to some of these private sector issues later in this chapter, in relation to South Africa.

PAWS did, in time, respond to many of the concerns raised in this section. In December 2004, Paul Turner of Northumbrian Water was recruited to the Secretariat as PAWS Manager (Clarke 2005; PAWS

2004d). Twelvetrees (2005) explained that the addition of Turner to the Secretariat was because PAWS needed 'somebody with a lot of experience to steer the partnership quite strongly . . . to make it more business like' and ensure 'that we were getting outputs and that we were delivering.' PAWS also recruited Ashley Roe, a private sector participant, formerly with Severn Trent Water International, to the Chair of PAWS. Turner (2005) reported that PAWS members 'decided that, from a governance point of view, we needed to have an independent chairman.' In April 2007, the Secretariat also moved home, no longer being hosted by DEFRA and instead being housed at Loughborough University, at which point a new non-DEFRA Head of Secretariat was also appointed from within the partnership (PAWS 2007).

Overview of the Southern Context

Having discussed PAWS in its northern context, we turn to its southern context – i.e. its operations in Africa. Unlike EUWI, PAWS has no formal relationship at a regional or continental level, but instead initiates partnership between northern and southern PAWS stakeholders. The logic underlying PAWS is that it should be demand-driven and therefore fully inclusive of southern partners. As Peacock (2003) explained, PAWS is there 'to offer our experience and skills to help meet the needs of partner countries that they have identified as their priorities.' Davitt (2004) similarly stated that 'the priority [for PAWS] is understanding need within beneficiary countries and working out ways of meeting that.' Although PAWS has now begun work in Ethiopia, our discussion in this chapter concentrates on the work it undertook in the three initial focal countries chosen in 2001, before the WSSD: South Africa, Uganda and Nigeria.

The first question to be addressed is how were these countries chosen as partners in the PAWS initiative? Peacock (2003) observed that 'there was a shortlist' of countries, but that he had 'no idea how that shortlist was reached' or 'the motivations' behind it. However, he suggested that 'South Africa was [chosen] . . . because they were hosting the World Summit, and the South African Minister, Ronnie Kasrils, was very supportive of it' (Peacock 2003). Similarly, an NGO partner representative (Interviewee 1, 2003) stated that, 'South Africa was a prime target . . . because the summit was in South Africa and because they have a radical, clear water policy . . . and there is some appetite for . . . [the] partnerships concept.' Peacock (2003) explained that Uganda and Nigeria 'were basically first in the queue . . . [having] said that they would like to be involved' (Peacock 2003).

However, some PAWS participants implied that there were other motivations behind the country choices. For example, Barker (2004) claimed that 'the decision for PAWS to work in Nigeria, South Africa and Uganda was to some extent driven by the fact that they were DFID focal countries.' Interviewee 1 (2003) agreed that 'it needed to be countries where Britain had positive relationships, had programmes,' adding that 'It was also where I think the private sector felt they had expertise and were already working or would be interested in working . . . and where things were happening in the sector where partnerships might be appreciated and valued as one of the answers.' Thus, it would seem, as Mitchell (2003) suggested, that the countries were 'decided on by a combination of where partners were interested in working, and where there appeared to be a demand for PAWS services.' Nevertheless, some interviewees felt that, as a result, PAWS must be classified as a northern partnership working in the south, rather than a north/south partnership. For instance, in April 2004, an NGO member insisted that 'It's absolutely a northern partnership . . . essentially it's a northern group going to a predetermined southern spot and fishing around for some work to do' (Barker 2004). A private sector member (Interviewee 14, 2004) agreed, stating that, although the 'aim has got to be' to have a north/south partnership, at the moment 'It's the north working in the south much more.'

Nonetheless, there was a recognition that while PAWS may have begun as northern-dominated, it needed to develop as a north-south partnership. For instance, Mitchell (2003) held that:

> In its formulation it was driven by a need here in the UK to develop partnerships to be seen to be supporting that aspect of UN operations . . . [but] what will ensure its longevity is if the demand and response from South Africa, Uganda, and Nigeria is that this really adds value . . . going from that kind of slightly supply-led approach to a very demand-led approach is what would give it wings . . . Otherwise it's just more patronising efforts from the north to tell them what we think they need.

PAWS sought to secure partnership agreements with the central governments of its three focal countries before moving on to work with other actors. As we saw in chapter three, the synchronization of the work of TTPs with southern governments' strategies is essential in securing any lasting and sustainable reform. One NGO interviewee (1, 2003) explained that 'There is a real challenge to support national government prioritization and their own development plans . . . analysis [has shown] that putting lots and lots of projects down on the ground, and supporting the sector or projects exclusively, has not done anything to create the sustainability and the prioritisation of that sector.' Therefore, it has become much more common for aid agencies to work through government, rather than by-pass it by doing direct project work, because when the aid agencies leave, it is indigent governments that have to ensure that their people's needs continue to be met. Mitchell (2003) stated that DFID 'believe the future is to fund through government rather than around government, otherwise you never build the capacity of government to govern . . . it's up to governments to meet the needs of their population, that's what they're elected to do.' Therefore, it was essential that the central governments in PAWS's focal countries were approached from the outset.

Note, however, that this was a government-to-government approach, described by Interviewee 3 (2003) as a 'a top-down initiative.' Rouse (2003) defended this approach: 'the great benefit of PAWS is that a government (the UK government) is working on that political commitment because otherwise it's quite difficult [to obtain] . . . for NGOs, and even more difficult for the private sector.' Likewise, Causer (2005) stated that 'we wouldn't get that support [from African governments] if we didn't have the UK government fronting it up . . . so it needs to be government led . . . we wouldn't be anywhere at all without a UK government lead.' Indigent local government was also engaged by PAWS, with much of the partnership's efforts directed at this level, working 'through the local government channels' (Oudyn 2004).

But what about the indigenous private sector and civil society? It was felt that such stakeholders had to be co-ordinated with indigent local government. For instance, an NGO partner (Interviewee 1, 2003) claimed that:

> most stakeholder approaches are interesting [and useful] if they enhance democratic processes but don't act as alternates . . .

> [because] local government is the first tier of representation, so local NGOs, local civil society, have to consciously work within the formal tiers of government because otherwise they're undermining what they're also trying to achieve, which is more democratic, more accountable institutions that are responsive to people.

However, although Mitchell (2003) conceded that 'at the end of the day provision [of water and sanitation services] has to come through the government,' he insisted that 'there is a very strong role for civil society and NGOs in empowering communities to demand services from their government.' Similarly, Rouse (2003) argued that:

> the top-down approach . . . on its own is not enough, you've got to have bottom-up [as well] because if you're going to achieve [the MDGs] . . . it can only be done by harnessing the energies of communities . . . you have to get their commitment, their ownership, or it's never going to be sustainable . . . in Europe and North America, it wasn't governments that originally determined water and sanitation, it was communities, and communities took responsibility for themselves . . . that's the way it happened, and that's the way it's going to happen anywhere else.

Rouse (2003) claimed that 'PAWS is trying to . . . harness the energy of the communities,' and in official documentation, PAWS declared that it was 'committed to the involvement of all stakeholders in partner countries particularly in the communities in which the partnership is engaged' (PAWS 2004b, 6). In line with this claim, an NGO partner representative (Interviewee 1, 2003) maintained that PAWS was trying hard to engage with indigenous civil society: 'that's certainly what their scoping studies have been trying to do.' Even so, at least for the first three years of PAWS's existence, repeated questions were raised about the extent of PAWS's engagement with civil society actors in the south, as we will see in the following country analyses. As Peacock (2003) said, 'The idea (which has not actually been realized) . . . when we get involved with the partner countries is to encourage them to have a similar base of government, private sector, and NGOs,' but when 'we tried to get out into the communities and talk to them about what they do and how they do it . . . I don't think there's been anywhere where the community has identified anything it wanted.'

Nowak (2003) was optimistic that when 'partner countries . . . see the [example of the UK model in the] way that government, the private sector and NGOs can work together in quite a co-operative way [it may be] . . . eye-opening to some people and . . . encourage similar groupings to come together in partner countries.' On the other hand, Davitt (2004) warned against forcing a multi-stakeholder model on African partner countries. Instead, he urged that there was a need to appreciate 'their culture and ways of working to understand how their society would view that . . . we shouldn't impose our model of what we think will work on what they need.'

Bearing the above issues in mind, the following three sections analyse how work has been conducted between UK stakeholders and the PAWS partner countries of South Africa, Uganda, and Nigeria. Specific attention will be given to our case study country of South Africa, where interviews were conducted in September 2005 with actors involved with PAWS.

South Africa

Our fieldwork was carried out in South Africa in 2005, because, compared with Uganda and Nigeria, South Africa had progressed farthest in its engagement with PAWS.

Introduction

Unlike Uganda, Nigeria, and the many other sub-Saharan African countries, analysts have long suggested that South Africa would easily meet the water MDG target (Interviewee 21, 2005; Khambule 2005; Quibell 2005; SANGOCO 2005; Weston 2005). Indeed, in July 2006, Gadebe (2006) reported that the Social Development Minister in South Africa, Zola Skweyiya, had announced that South Africa had already achieved the Millennium Development Goal' on water, way 'ahead of schedule.' However, the South African government took things a step further and 'committed itself to ambitious targets to do much better than the MDGs with everyone having access to water by 2008 and everyone having sanitation by 2010' (Interviewee 21, 2005). Therefore, South Africa would be near the bottom of any list which ranked African countries by their need for external assistance on water supply related issues. As a result, Waller (2003) said that 'South African water services are completely different from the rest of Africa.'

This means that as a case study, South Africa is somewhat unrepresentative. However, every country's problems and issues are context-specific, and the main issue PAWS tried to address in South Africa – capacity building in municipalities – is not an unusual area of weakness for sub-Saharan African countries, although in other countries this is usually but one of a vast array of problems. Furthermore, while South Africa is not 'a poor third world country . . . there are parts of it that are,' and there are problems for which it welcomed support from PAWS (Interviewee 1, 2003). Indeed, according to a 2004 report by the UNICEF and WHO Joint Monitoring Programme on Water and Sanitation (JMP), entitled *Meeting the MDG Drinking Water and Sanitation Target: A Mid-Term Assessment of Progress*, based on data up until 2002 (WHO/UNICEF JMP 2004), while South Africa was on track to meet the water MDG target, it was not on track to meet the sanitation target. What is more, a representative from the Mvula Trust (the largest water and sanitation NGO in South Africa) (Interviewee 21, 2005) reported in September 2005 that in reality, even if the internationally agreed sanitation target was met in South Africa, along with the already achieved water target, 'Very few believe' the government 'will meet its own commitments on water and sanitation' because they are 'not on track at the moment.'

Moreover, Mitchell (2003) asserted that the progress that had been made by South Africa in improving water and sanitation provision might have been made too quickly to be sustainable. As a result, he argued that, 'they're probably storing up some problems for themselves in the future . . . [and] PAWS can really help with that' (Mitchell 2003). Derek Weston (2005) of the South African Department of Water Affairs and Forestry (DWAF) admitted that 'I'm worried about the sustainability of many of the things we're doing,' while Gavin Quibell (2005), a Technical Support Consultant for DFID in South Africa, acknowledged that, while:

> we have posters up saying we've provided 13 million people with water, it's another thing entirely to say that that water is sustainably provided. Indications are that in many cases it's not . . . any water supply and sanitation scheme requires some maintenance . . . in some cases operation and maintenance is not as effective as

> it should be . . . the focus is on providing new stuff rather than fixing the old.

Before the end of apartheid and the first democratic elections in 1994, there was no central government department in South Africa that held responsibility for the delivery of water supply and sanitation. As Muller (2002) explained:

> Responsibility was fragmented and allocated to local governments in the previous four provinces and to the ten nominally autonomous homelands, resulting in different levels of service. Most of the then white local governments offered standards equal to those in industrialised countries. In the rural areas there were often no services, while in black urban areas the situation was mixed. Both urban and rural services for black people were often in a state of disrepair. This situation was exacerbated by the absence of any coherent national policies, guidelines or support structures . . . In 1994, the new government made DWAF responsible for ensuring that all South Africans have equitable access to water supply and sanitation.

Water law was completely reformed by the 1997 Water Services Act and the 1998 National Water Act, and responsibility for water supply and sanitation was devolved to newly structured local government authorities. However, as Quibell (2005), explained, 'in South Africa our ability to plan things, write legislation, write policy, far exceeds our ability to actually implement it . . . this does not mean that we are bad implementers, rather that we tend to write policies and legislation that are very difficult and expensive to implement.' The problem was (as in much of Africa) that local government was 'grossly under-resourced' (Interviewee 1, 2003), and often lacked the 'skills and capabilities to run the water and sanitation services' (Waller 2003). As Quibell (2005) put it, 'local government is key . . . but unfortunately they are very under-capacitated at the moment . . . the smaller towns can struggle to get in the right kind of skills.' Similarly, Christo Nel (2005), Head of Technical Services for the Zululand District Municipality (ZDM), stated that:

> We've got a major problem now in the country with the transformation that has taken place. The environment has been created for

> previously disadvantaged individuals to actually get into positions where they were previously not able to participate. The result is that you get many, many posts being filled by people who do not have the capacity and experience to actually do the job.

Capacity building at the local level was, therefore, the main area identified by PAWS where it could offer assistance. In South Africa, PAWS did not create new programmes of work, but supported existing programmes and helped build capacity in-country, leading one interviewee (Khambule 2005) to remark that, 'it's seen as a vehicle to make things quicker, faster, better.' Alfred Legoabe (2005), a Contract and Legal Services Manager with the South African central government Department for Provincial and Local Government (DPLG) explained that 'particularly in the municipalities we have a dearth of skills . . . the fact that PAWS will actually bring in the expertise . . . [and] capacity building . . . is absolutely vital . . . If you haven't got the skills to utilize effectively . . . the little money resources that you have, then you are in all sorts of trouble.' Therefore, the PAWS approach in South Africa was needs-based: as Davitt (2004) said, 'it's been very much locally based around their priorities.'

Although by 2005, out of PAWS's three focal countries, most progress had been made in South Africa, the process in-country was difficult, and the initiative took a long time to gain any sort of momentum. As Mitchell (2003) reported, 'it took over a year just to get to a position where we'd identified the municipalities we were going to work with.' He suggested that this delay was because PAWS chose to work through central government rather than around it: 'we had to first get their trust in the process' because the 'central government . . . has a certain level of veto over what the municipalities do,' but 'working through the national government bureaucracy really is time consuming' (Mitchell 2003). Causer (2005) pointed out that the role of the South African central government was also important in evangelising and mustering up support for PAWS at lower levels: 'no-one would be interested unless the message had come down from the top in South Africa that they wanted PAWS to succeed.' Legoabe (2005) explained that 'it is important for DPLG to put its weight behind PAWS because . . . municipalities . . . in most cases, respect the name of the DPLG . . . [without DPLG support] you would probably have a situation where a number of municipalities would not have much interest in the whole thing.' Thulise Khambule (2005), who was appointed as the South Africa PAWS co-ordinator in September 2003, suggested that the fact that there were people working in DWAF, SALGA (the South African Local Government Association) and DPLG on PAWS, 'is a demonstration of commitment . . . people are accountable on it, they report on it' so 'it's not like an external thing,' it had some degree of in-country ownership.

However, DPLG were not initially involved: they were brought on board at a later stage. At first, DWAF were asked to identify the municipalities it would like PAWS to work with, 'but they didn't get anywhere fast with DWAF' (Peacock 2003). Instead, SALGA (which was 'mandated by the new South African Constitution to assist in the wholesale transformation of local government in South Africa from the pre-1994 regime to the new dispensation under the country's first democratically elected government' (SALGA 2006)), 'stepped in and organized a bi-national meeting at which . . . [PAWS-UK] were presented with four municipalities to work with' – Nkomazi, uThungulu, Matjhabeng and Zululand (Peacock 2003). At the beginning of 2003, scoping visits were undertaken to each of these four municipalities, where small teams from the UK 'went out . . . [and] learnt what their needs and priorities were.' Nel (2005) reported that 'there were no predetermined areas of support' that PAWS designated, and although this made the purpose of PAWS somewhat unclear at the outset, it 'was a good approach because during the process there was interaction.' In Zululand, for example, Nel (2005) stated that this meant that PAWS concentrated on 'issues that we have identified . . . [such as] the metering of water, water quality, asset management, training, [and] customer care.' Following those visits, 'reports went to African municipalities and UK partners,' and at a UK Forum meeting in May 2003, the Secretariat said 'right we've got their needs, we know what they want – are we able to help them?' (Peacock 2003). The response was very positive, and teams were set up to work in several areas. In June 2003, a bi-national meeting was held in South Africa, where UK partners from DEFRA, DFID, Thames Water, Severn Trent Water, and the trade union, UNISON, met with municipality representatives and officials from DWAF and SALGA, to discuss the outcomes of the scoping studies and consider a way forward.

However, work did not develop at a great pace. As a private sector respondent (Interviewee 14, 2004) explained in April 2004:

> there is a need in South Africa but it is proving very difficult to make any inroads . . . South Africa is a country that doesn't really need our technical assistance . . . The problems in South Africa . . . [are] institutional . . . they've got the drilling specialists, they've got the hydro-geologists, they've got all that help, it's institutional help they need.

Assistance with institutional change is 'about influencing, it's about political lobbying . . . it's . . . long-term' help, and is a much 'slower process' than technical assistance (Interviewee 14, 2004). An NGO representative (Interviewee 1, 2003) said that PAWS became 'almost like a morale boost, a confidence builder,' helping with 'competence building.' An example of the type of assistance given by PAWS was described by Nel (2005) as 'customer care': 'customer care . . . is something we've never had in the country . . . we are totally incompetent as far as that's concerned,' so a group of South Africans visited the UK in 2004 'to see how the customer care centres are set up' and where the UK partners 'shared their experiences . . . The aim is not to get somebody who can show you how to do it and leave, the aim is to build the capacity in-house.'

PAWS did not, however, have any financial resources to offer to recipient or partner countries, and this was an issue of concern raised by some southern participants. For example, Legoabe (2005) suggested that when it was 'made clear that this is not about funding, this is about intellectual assistance . . . a lot of municipalities kind of raised an eyebrow . . . if it is at all possible for some funding to be made available then that would be good . . . Who would say no to money?' Similarly, Quibell (2005) stated that 'financially it's too small to make a huge difference.' However, Nel (2005) disagreed, maintaining that:

> Funding will . . . not solve the problem because we are here forever and we need to build our own capacity . . . We do need . . . a lot of funding but you can put a lot of infrastructure into the ground, then you have to operate and maintain it, so the managing of infrastructure needs skills . . . you cannot buy experience, you cannot buy knowledge.

So the fact that PAWS did not offer any funding did not fatally weaken its impact.

Programmes of Work and Obstacles to Progress

That the initiative chose to work first with central government (before moving on to work with lower tiers of representation), and that the type of assistance required in South Africa was institutional support and capacity building, were not, however, the only reasons for slow progress in-country. Davitt (2004) reported that the initiative 'found great difficulty in making significant progress' with the municipalities, and Peacock (2003) suggested that PAWS had come up against many 'political and organizational hurdles' within the municipalities. For example, Peacock (2003) reported that in some cases, the UK partners believed they had agreed on a programme with a certain municipality, only to find that the programme 'has to go through a committee system . . . [and] some of the committees only meet every two months or so,' which demonstrated that political endorsement at the municipality level could be just as cumbersome to obtain as at the national level (Peacock 2003).

Moreover, the initiative suffered from considerable difficulties in maintaining communications between UK stakeholders and South African participants, because 'trying to form an effective communications link is hard,' and 'you always have that vulnerability of it coming down to individuals' (Davitt 2004). Some of these problems were due to inappropriate communication methods. As Clarke (2005) pointed out:

> one of the problems we've had in PAWS is that we send things over by email [to South Africa], and the fact you don't hear from them isn't because they don't think it's valuable, it's because they never got it because their email crashed. So email is often not the best way of talking to people which is something we're learning . . . telephone and fax are much more useful.

Communication difficulties are linked to the capacity-deficit problem – many of the South African actors did not have the capacity to fully accept PAWS's help and maximize its potential. For example, at the national level, Peacock (2003) cited an instance where DWAF asked for help in creating a Customer Care Charter Model (CCCM). South East Water then 'produced lots of information and materials which basically, when put together, made up a CCCM. Then PAWS heard nothing again. The problem is a need for someone to contact over there and there isn't a contact.' At the municipal level, a private sector partner (Interviewee 14, 2004) gave the example of the problems encountered with one of the municipalities involved, which he described as 'typical':

> [There was a] scoping visit in 2003 . . . and the people we met were very receptive and we came away with a lot of ideas on how we could improve things and help them to build capacity . . . it was all about health awareness, general education, use of staff, communication to the customers, a whole range of these sorts of things . . . and then nothing happened at all. We went back out . . . almost a year later, we renewed the relationships we'd made with some of the key players and we went . . . to focus on the areas that we'd tried before . . . Again we gave the report to the council, to the Chief Exec, the Municipal Manager of the council, and that's it, we've walked away . . . All of us have tried to maintain communication [with the South African actors] but there is only one email address for the whole council . . . the guys that we were working with, the practitioners, have no, or limited access to the internet . . . it's great when you're there, you're working hard and you're talking to the guys, you're joining things up. As soon as you come away then they go back to plate balancing and solving the day-to-day problems . . . so it's very frustrating . . . there's so much difficulty with communication . . . but if nothing is going to happen and they want to do things at their own pace, then fine, we're here, we're available, but let's move on with the partnership to where we can have more value.

Davitt (2004) agreed: 'Part of the issue around capacity building . . . is . . . to have sufficient resources within the municipalities to define what they need and receive support . . . if the capacity isn't there, how do you get in to build it if it's not there to build on?' This capacity deficit included a lack of experience: as Nel (2005) of the ZDM, pointed out, 'the capacity of the officials in local government to participate' in PAWS must be present for the 'PAWS programme to be successful . . . you cannot help people that do not have the basics and at least a certain level of experience . . . Some of the local government partners that have participated just do not have the people to understand and actually communicate with the [UK] partners.'

Of the four municipalities that PAWS initially engaged with, Turner, speaking in June 2005, contended that 'the successes have been with the KwaZulu-Natal ones, uThungulu and Zululand . . . those two are working fine.' Work here was primarily led by Yorkshire Water on the UK side, and there 'is a good working relationship' between them and the municipalities, 'but they're the ones who are probably closer to the UK in terms of the level of the sophistication that they have' (Turner 2005). Turner (2005) said that 'the two districts that need the most help . . . are Matjhabeng and Nkomazi,' but 'they haven't actually got the capacity to take our help and if we can't maintain a dialogue with them, then there isn't much point in us actually working there.' This suggested a catch-22 situation, in that the PAWS initiative was aimed at helping to build the capacity of South African municipalities, yet the municipalities' lack of capacity was a barrier to accepting this help. As a result, Turner declared in June 2005 that:

> we're not walking away from South Africa but we are getting to a point now where we're going to be more measured with the effort that we put in. We like to think of PAWS as being demand driven, so that goes for partners and for work. Once they know about us and we've explained what PAWS can offer, if they don't want that support then fine, we'll walk away because there is no point us pushing too far.

In September 2005, Legoabe (2005) stated that the capacity of municipalities to receive help was a problem that the South African actors themselves had to solve:

> In South Africa we are still a little bit slow to come up to a level where we can fully engage PAWS and have all the benefits that PAWS can give us . . . it depends on us to make PAWS have much more of an impact than it does now, we must fully engage PAWS. We have to get complete buy-in from the municipalities, they must fully prepare themselves for PAWS injection, and I'm confident that we can do that.

Although Nel (2005) said that it was a problem that PAWS did help with 'through seminars . . . [and] working sessions' – for example, the South Africans who went on the visit to the UK in 2004 'actually split up into groups that had certain levels of capacity, and the groups with the least capacity went for an intensive training session . . . so the guys came back with much more knowledge of how things are to be done' – he acknowledged that while such assistance could help in a small way, the general capacity deficit could not be addressed in this manner: 'responsibility lies within the country, with the Department of Local Government, with the Department of Water Affairs, they need to set up structures and mechanisms to do training, to get people's capacity sorted out.'

There were also organizational and political barriers to effective co-ordination and communication between South African participants, some of them caused, or at least compounded, by PAWS-UK. For instance, when PAWS-UK approached SALGA to identify the four pilot municipalities with which they could work, this created problems for the partnership, in that DWAF and DPLG felt sidelined; 'it was perceived by DWAF and DPLG as a SALGA thing' (Khambule 2005). However, this problem was overcome with the creation of a National Steering Group in early 2005, made up of representatives from DWAF, DPLG and SALGA, as well as the PAWS co-ordinator, Thulise Khambule, who was appointed in September 2003. The recruitment of a national co-ordinator for South Africa also addressed other problems with PAWS. Khambule, who was previously employed by DFID in South Africa, was appointed to spend 70 per cent of her time working for DFID and 30 per cent working for PAWS. Khambule (2005) explained that she was 'like the eyes and the ears of the UK partners,' and that as well as working 'closely with the Secretariat,' her 'role is also to assist the South African partners because they don't have Secretariat support.' This, Khambule (2005) stated, involved supporting not only members of the PAWS Steering Group in South Africa but also:

> local partners as well, the municipalities . . . helping and facilitating the discussion and making sure there is political buy-in, getting them to understand what the partnership is all about, what it is trying to achieve and how we expect the partnership to work in the municipality and with the UK partners through the Secretariat . . . making them aware . . . encouraging them to participate, and identifying with them the benefit of participating

> in the partnership. Then taking that participation, commitment and willingness and translating it into a Memorandum of Understanding . . . The other thing is an identification of needs . . . we have to sit with them and develop a programme with them to say, ok, if this is an issue that you have identified, how can the UK partners assist you in addressing this?

A private sector participant (Interviewee 14, 2004) noted of Khambule's contribution to PAWS that 'it's ideal because then you've got a local contact who knows how it works on the ground.' Similarly, Causer (2005) said that:

> Thuli is vital to our activities . . . she . . . makes things happen, she's a key player . . . And of course she gives us credibility . . . if we just turned up, five white persons from London, a black audience might dismiss us, but the fact that . . . we have Thuli as an integral part of our team and it's clearly a very tight relationship . . . gives us credibility with the black representatives.

Legoabe (2005) also stated his confidence in her abilities, asserting that 'Thuli is very, very proficient in that job . . . Almost alone Thuli can drive this process, particularly because she is aware of the fact that her resources are not necessarily in her office . . . I am part of her resources as are the rest of the government departments . . . I am sure she is the right person to coordinate.' Similarly, Ashwin Seetal (2005) acknowledged that:

> Thuli has basically been my mainstay in this whole business, she's a first class resource to have around . . . it takes the administrative burden [away from me], the linkages, facilitating, that kind of stuff . . . apart from taking care of the logistical arrangements . . . Thuli's advantage is she understands a lot of the water services stuff and she picks up the water resources stuff very easily.

Indeed, Seetal (2005) argued that such a co-ordinator was needed in all of PAWS's partner countries if the initiative was to be successful: 'in each of these countries, not just South Africa, you need somebody there, preferably a local person, driving it in that country' – i.e. to give it anchorage.

At the end of 2004, it was decided that PAWS should expand its capacity building work to other municipalities, beyond the original four. Yet, although PAWS was now talking about the necessity of 'capacity to receive', and had Khambule on board, the initiative still encountered problems in identifying municipalities to engage with, and in building relationships with them (Khambule 2005). Legoabe (2005) of DPLG, explained that, when representatives from DPLG and DWAF met with Khambule in January 2005 to put together a business plan for PAWS:

> the one component we couldn't finalize was the involvement of the municipalities because as much as we identified the municipalities that we felt were in dire need of PAWS input, we also had to get the green light from those municipalities as well – for them to say we want to come on board and benefit from PAWS input.

Indeed, it took until April 2005 for the list to be finalized (PAWS 2005b, Annex E). However, Khambule (2005) contended that the main difficulties encountered in the municipality identification process came about, not because of the municipalities themselves, but because there 'was a problem with the roles of the South African national partners' in determining whose responsibility it was to put forward a 'final list of municipalities, to say these are the municipalities that we are selecting for the second round . . . with the new five it took time between DPLG and DWAF to agree on that list . . . DWAF and DPLG have different priorities.' These five municipalities were Sekhukhune, Mareng, Mogale City, Ndlambe, and Central.

Once the five municipalities were selected, their representatives were invited to a workshop to present their needs. However, as Legoabe (2005) reported, 'Of the five only about three turned up.' Khambule (2005) insisted that this was not due to a lack of commitment from the municipalities to PAWS:

> otherwise they wouldn't sign a Memorandum of Understanding . . . we've learnt from the previous four municipalities, we know what we did wrong, what we should have done to make things faster and better and now we're taking those lessons into these new ones and saying immediately we need to have MoUs so that both parties commit to this.

Legoabe (2005) said the problem was, again, a lack of capacity within the municipalities:

> I can give you the example of one municipality . . . I . . . met with the municipal manager [and others, to discuss a matter unrelated to PAWS] . . . and I raised the issue that I had tried on countless occasions to contact him [about PAWS] and he had never come back to me, and he had been difficult to locate . . . he said, no, no, we are very interested in PAWS . . . [and] you would expect in that kind of an environment this guy would be alarmed and that when I contacted him again he would respond immediately, otherwise he would suffer the same embarrassment in another meeting. I have his cell number, I have his office number, I have his fax number, yet I have not drawn a response from him . . . it is very hard to get hold of these people . . . the strategy we are following is the right one, it is just that the municipalities are very, very slow . . . we are holding PAWS-UK up, they are standing there waiting for us and we are moving very, very slowly . . . it is clearly an indication of a lack of skills.

Thus a lack of capacity to receive assistance was clearly a large stumbling block for PAWS. However, Khambule (2005) claimed that this was not something that PAWS could hope to address in any substantial way: 'it's an internal issue, the capacity issue is a South African issue that they need to work on and prioritize.' In other words, while PAWS could help with building capacity to deal with specific problems such as billing or customer care, it could not seek to tackle the general capacity deficit that existed in local government because this could only be overcome over a relatively long period of time.

On the other hand, having a highly proficient co-ordinator on the ground who could facilitate communications and develop the programme, undoubtedly assisted PAWS in its work in South Africa. Khambule (2005) reported that since she joined PAWS, there was less fractiousness between South African national partners, and greater co-ordination between different levels of the PAWS structure in-country. For example, whereas in the first four municipalities PAWS worked with, DWAF regional offices were not involved (perhaps because DWAF had not played a significant role in choosing these municipalities and felt marginalized), in the five new municipalities Khambule (2005) stated that 'each time I go to a municipality I take the (DWAF) regional director with me to that meeting, so from national to province to the local, there is a direct line of communication.'

As well as working at the local level, PAWS also undertook work at the national level in South Africa and provided 'some strategic support . . . in the form of regulatory advice, [including] an inward visit to the UK' (Turner 2005). PAWS has also conducted a series of seminars on topics such as contract management and organizational change, which were delivered to municipalities beyond the four directly involved in the initiative. These were very well received, and Davitt (2004) argued that at this level, PAWS was successful: 'the national seminars and guidance on a national level on the framework for institutions . . . is working quite well.' The only concern here was, as Peacock (2003) explained, about 'how much the seminars impact on practices.'

Non-Governmental Stakeholder Participation

Criticism has been levelled at the PAWS initiative on two distinct, but related, levels with respect to non-governmental stakeholder participation (including both the not-for-profit sector and the private sector) in its work in South Africa. First, indigenous non-governmental stakeholder participation in PAWS's work was virtually non-existent. Second, while UK private sector partners were heavily involved in PAWS's work in South Africa, there was minimal engagement by UK based NGOs, other than the trade union, UNISON.

The PAWS *Memorandum of Understanding with South Africa* between PAWS-UK and the South African Government explicitly stated that 'The Government of South Africa supports a partnership between Government, civil society and the private sector in South Africa to engage with the UK based partnership,' and that 'It will be the intention to form local project groups at the municipality level, which will aim to be representative of the three sectors in the local community' (PAWS 2004b, 23). Thus, PAWS-UK clearly regarded southern governmental commitment to the participation of the south's private sector and civil society as essential. Moreover, the South African government agreed with this emphasis.

However, in its first five years of existence, PAWS failed to step significantly beyond engagement with government (national and local) in South Africa, to involve stakeholders outside the governmental sphere.

Waller claimed in November 2003 that 'On scoping visits they have talked to communities, but that is just more of a fact-finding exercise to understand the issues.' In essence, 'They haven't got past talking to governments.' Similarly, Turner (2005) acknowledged that 'In South Africa we've never been able to make proper engagement with civil society, it's been very much with government, and the private sector input has been very much coming from the UK partners.' Likewise, Davitt (2004) admitted that in the scoping visits, 'in the municipalities that we're trying to link up with we've just been talking to the local government offices, rather than any other representatives of civil society.' Another private sector participant (Interviewee 14, 2004) reported that PAWS was 'working exclusively with either municipalities or central government' and that there had 'been no involvement of any NGOs in South Africa.' Causer (2005) noted that 'to the extent that there are private sector players in South Africa we've not drawn them into the PAWS partnership.' So PAWS did not succeed in engaging with either civil society or the private sector in South Africa.

PAWS did try 'to have a relationship with the Mvula Trust, a major national water and sanitation NGO in South Africa' (Interviewee 2, 2003), but that link proved problematic, because, as Peacock (2003) explained, the Mvula Trust was 'almost like a private sector company,' and it raised the familiar issue of using PAWS as a means to gain business contracts:

> we're prepared to work with anyone . . . if they have a contribution to make and they're not doing it for purposes that don't fit in with the way that the partnership works. We have no objection to NGOs in South Africa who expect to get contracts for things, but I think we need to be a bit careful about how we get involved with them and what they then do with that experience that they've got from us.

The problem was that PAWS's acceptance of Mvula as a partner might have undermined the government's role as a provider of services. However, a representative of the Mvula Trust (Interviewee 21, 2005) argued that the fact that many South African NGOs operated (like Mvula) as deliverers of services, as opposed to campaign or advocacy organizations, was a result of northern donors' refusal to allocate to them funding for civil society participation. This interviewee (21, 2005) contended that 'the EU is the one pushing hardest for civil society to be more involved . . . but they put all their money into the government budget.' This meant effectively, that the EU and other donors were turning 'NGOs into non-profit consultancies' because of the NGOs' need to access funding in order to survive (Interviewee 21, 2005). On the other hand, if donors fund NGOs directly, this can undermine indigenous government institutions. So TTPs have to walk a fine tightrope between supporting the development of a strong civil society, the establishment of which would help strengthen governance as a whole, and avoiding the creation of parallel power structures that could weaken government's ability to fulfil its responsibilities.

Admittedly, as largely implementers and service deliverers, Mvula Trust, and other organizations which operated on similar principles, could not be said to represent the views of communities, or be a voice for "civil society". Nevertheless, a representative of the Mvula Trust (Interviewee 21, 2005), while admitting that 'not being a membership organization . . . we don't claim to [represent civil society within South Africa, and] we don't have a mandate to,' pointed out that the Mvula Trust 'do work with hundreds of communities throughout the country . . . [and] have a good sense of what is going on . . . and what people need,' and in this sense, were equipped to 'articulate fairly well on behalf of communities.' More generally, although Nel (2005) reported that at the national level, 'the unfortunate reality is that NGOs in South Africa are not really that strong [and] there is a general feeling that they also lack capacity,' a representative of the Mvula Trust (Interviewee 21, 2005) noted that there were 'plenty of CBOs [community-based organizations]' and civil society groupings at the local level, which PAWS could engage with.

Another obstacle to non-governmental stakeholder participation in South Africa was that it was mediated through local government. Khambule (2005) noted that in South Africa, 'the constitution gives the responsibility to government to ensure public participation in decision-making processes. So that is a high-level commitment by the South African government.' Legoabe (2005) argued that the key role in carrying out this responsibility was played by local councils:

> The municipalities in the implementation of projects are compelled, in terms of legislation in South Africa, to interact with the

> communities. In other words, even before a municipality would engage somebody from elsewhere to come in and undertake a project in their area, the municipality is compelled to get council approval for that, and the council is compelled to inform the community about the fact that PAWS is coming to that community . . . So the community has to give a blessing to that, and not only the community, the labour movement has to give a blessing to that as well. That is legislated and the municipality will not go any other way. So the community has a way in which it makes input into the whole initiative. It might not be as conspicuous as it is in the UK, but civil society is involved . . . through the municipalities.

However, it was not clear to what extent the community was informed by the local council of such activities, let alone genuinely consulted. Furthermore, Legoabe's comments made it clear that South African communities could in no way be considered as partners in PAWS, even though they might be granted some input via the municipalities. Therefore, as an NGO member (Interviewee 3, 2003) argued, the UK-based PAWS partnership 'probably could have done more to involve communities but it does work within the constraints of what partner governments want to do.'

Some interviewees suggested that the degree of importance ascribed to the engagement of indigenous civil society actors in Africa rested on what it was that PAWS aimed to achieve. A not-for-profit partner (Interviewee 2, 2003) maintained that 'PAWS is, by and large, about bringing in technical [or institutional] support, whether on the regulatory side, the engineering side . . . or the construction side of things,' and therefore, 'it is more important that civil society is engaged by . . . [African] stakeholder groupings rather than [by] PAWS.' This interviewee (2, 2003) added that 'PAWS can't be expected to be the liaison point for all different actors in South Africa, that's not its function . . . I see it as the South African partners' role to bring in civil society.' Nel (2005) agreed, stating that, while PAWS's work should be multi-stakeholder within South Africa:

> it cannot be done by international partners, they can facilitate the process but you still have to have the initiative from within the

> country, for the people to stand up and say 'let's do it', otherwise it's not going to work. There needs to be a voluntary effort to participate and share ideas but the problem is that everybody is so busy that it's not a priority at this stage.

Khambule (2005) similarly contended that non-governmental participation in PAWS is 'an issue that needs to be dealt with internally . . . there are some initiatives around civil society participation in government programmes and it's done through DPLG, it's done through SALGA, it's done through DWAF, so I think gradually it's going to happen.'

Another factor inhibiting civil society engagement was the perception by some that PAWS was essentially offering free consultancy services. Causer (2005) stated that 'A lot of what PAWS does is effectively consultancy work.' Legoabe (2005) said that 'if I engage a consultant then I would have to pay that consultant, whereas if I get a PAWS person to do that I don't have to pay.' Waller (2003) claimed that if this was, indeed, all that PAWS aspired to do, then it made the need for a partnership redundant: 'They could have had one to one conversations with the government and offered some free consultancy to particular agencies and they didn't need a whole PAWS business.' However, Khambule (2005) argued that there was a major difference between normal consultancy work, which was short-term, and the kind of help PAWS offered, which was long-term:

> if you send a consultant somewhere you pay the consultant, and after six months the project is finished and he goes and you are not sure whether there is ownership of whatever was done . . . but PAWS will be there in the sense that after six months or so they will come and visit you, check on you, you will correspond by email, you call each other. If you are a municipality that's your second office, your steady hand, that's where you draw resources from and gain confidence . . . it is not a one off thing.

Similarly, Nel (2005) argued that 'what PAWS does is sustainable . . . PAWS for me is about intellectual capacity, and [support is] . . . ongoing, it's not just going to stop.'

Another factor explaining why indigenous non-governmental participation in PAWS in South Africa was so limited was that UK NGOs such as Tearfund and WaterAid had effectively no involvement in PAWS in its South African work. This led some PAWS participants to suggest that the initiative had failed to fully utilize its partners' experience to forge links with local civil society groups. As one private sector partner (Interviewee 14, 2004) claimed, 'partnership . . . is about exploiting synergies between the partners, and [PAWS is not] displaying much synergy.' Peacock (2003) wanted the UK NGOs to be more fully engaged within South Africa stating that 'Ideally . . . [they] should be advising and helping and supporting on the community issues because we simply don't know when we go to these municipalities what we're going to be presented with' (Peacock 2003). WaterAid operate in a decentralized manner: they do not undertake project work from their London office, but have country offices with their own programmes of work. For its part, Tearfund do not have country offices, but work in partnership with local actors in specific countries to empower them to undertake advocacy work themselves. This means that while these two NGOs had 'more experience of working with communities than anybody else' (Rouse 2003), both WaterAid and Tearfund could only offer assistance to PAWS where they 'have specific resources that match up with what the initiative is looking for' (Interviewee 1, 2003). Neither of these NGOs worked on water and sanitation issues in South Africa, and of the three countries chosen, Tearfund only had partners in Uganda, while WaterAid only had offices in Nigeria and Uganda. So, as a private sector participant (Interviewee 14, 2004) pointed out, 'although there are four pilot projects . . . [in South Africa] they're not truly testing the partnership and the partnership principles because the only NGO involved . . . has been the [trade] union UNISON.' With regard to UNISON's engagement in South Africa, Peacock (2003) said that 'One of the things that we've insisted on is meeting the trade unions when we've been to each municipality and that's been useful.'

Water Resources Management

In 2004, PAWS began a new programme of work in South Africa, focusing on water resource management. The UK Environment Agency (EA) was heavily involved with this programme, offering advice, training and information on the 'tools', 'systems and procedures', 'methods and techniques' used in the UK 'for catchment management,' as well as offering assistance with the water allocation reform process in South Africa (Twelvetrees 2005). Although, initially, water supply and sanitation were the prime focal areas for PAWS, once the EA joined the partnership in early 2004 (Twelvetrees 2005), and could offer expertise in water resource management, it made sense to broaden its scope. Indeed, Ashwin Seetal (2005), Director of Water Allocation in DWAF, asserted that this development was necessary: 'PAWS has mainly been involved with water services but water resources need to be featured as part of the entire water management approach within the PAWS programme.' Similarly Clarke (2005) commented that:

> If you're talking about integrated water resource management, you can't talk about supplying water without considering the resources side as well, so they do fit together very well . . . in terms of it being a move in a different direction . . . it's just an area that we've had requests for help on and so we're responding to that, and the Environment Agency are able to provide that.

A mini-scoping visit was undertaken by a Secretariat member and an EA official in August 2004 to build the foundations for this work, and in April 2005 a fuller scoping visit was undertaken by a team of EA specialists. This trip entailed 'three intensive days of discussion' between DWAF and EA officials (Clarke 2005), and was aimed at seeing how PAWS could contribute to 'the development of the new Catchment [Management] Agencies' (CMAs) being set up in South Africa (Twelvetrees 2005). The ToR for this project (PAWS 2005e), stated that there were six deliverables: to

- Explore capacity building opportunities to enable DWAF to better determine, shape and facilitate the development of the CMAs . . .
- Explore opportunities for individual and institutional capacity building to enable more effective cooperative governance within DWAF and between DWAF and other government departments . . .
- Explore opportunities for joint institutional learning between emerging CMAs and operational arrangements in England & Wales for catchment management and regulation – e.g. through twinning . . .
- Explore capacity building opportunities to support Water Allocation Reform through more effective stakeholder engagement at the

municipal level (including the role of MoUs with local government institutions) . . .

- Explore capacity building opportunities to address cost issues and the social economy . . .
- Explore opportunities for input into the guidelines for developing Catchment Management Strategies . . .

Although 'A lot of the projects will just be a short-term sharing of information,' there were also plans 'to do . . . basin twinning' between the River Ribble Basin in the UK, which was a pilot testing ground for the European Water Framework Directive, and a basin in South Africa, possibly the Inkomati Basin, which was the first Catchment Management Area in South Africa to appoint a Catchment Management Board (Clarke 2005). In late 2005, a team from South Africa visited the UK to develop twinning arrangements, talk with EA staff in Bristol, and spend some time at the Ribble Catchment to see 'exactly what's going on, and how they're doing it on the ground' (Weston 2005).

In the work that PAWS carried out on Catchment Management, Twelvetrees (2005) reported that the northern partners were 'trying to push that stakeholders are relevant . . . so we hope to work with various stakeholders but we're taking the lead from DWAF.' However, Weston (2005) claimed that 'It's been very much focused around, not just DWAF and centralized government, but also the agencies. We've not really taken it one step down to civil society, stakeholder level. That probably comes from our side in that . . . it's institutional memory we're worried about.' Moreover, in relation to the work on water resources, some South African participants expressed concern about the partnership modalities implied by PAWS. For example, Seetal (2005) questioned how much genuine partnership existed between South African and UK actors, stating that:

> If I talk about partnership it's about people coming together, bringing something into the relationship that both can benefit from . . . the UK partner side are bringing a lot of financial resources as well as intellectual capital. From the South African side we can match them possibly on intellectual and experiential sharing, but not much on anything at all in the way of financial resources. So it becomes an unequal partnership . . . What is the

> point of the relationship? . . . what is it that people want within any programme, not just PAWS?

In other words, what were the motivations of the UK players in getting involved in the water resources management work of PAWS? Seetal (2005) claimed it was not clear:

> I think within PAWS you've got about 30-odd partners in the programme itself [in the UK], how do they come in, it's still not very clear to me . . . what does the institutional arrangement mean? . . . Is it political, is it a business venture under the guise of a support programme? . . . it's not clear to a lot of people . . . I like certainty, I like clarity on the specifics of what the programme has to offer, both to ourselves and to the British partners involved . . . One of my concerns is that that is not clear to anybody at this stage. I think at the strategic level the intentions of the programme need to be clearly spelt out . . . we have a lot of international partners in many of our programmes out here and having worked with a number of these over the years, it takes a lot of time and effort, so to bring another one on to the landscape doesn't make my life easy.

Seetal wanted assurances that the engagement by UK partners would be worthwhile for South Africa, and that their motivation was not to secure contracts for UK firms. The issues raised here by Seetal were addressed in PAWS's work in the municipalities on water supply and sanitation, by the creation of the PAWS *Rules of Engagement for Partners' Participation in Projects in South Africa* in June 2003. This document explicitly stated that 'Private sector involvement is not seen as a way for the UK based businesses to short track contract wins nor as a means of promoting the privatisation of water' (PAWS 2003g). These *Rules of Engagement* were drawn up 'to ensure that no conflict of interest arises if the possibility of commercial activity arises from the PAWS projects within that Project Municipality,' and it bound partners to 'agree that they or any subsidiary or associate will not participate in or engage in a tender process for any commercial activity within the "Project Municipality" for a period of 2 years after the end of the PAWS project within that "Project Municipality"' (PAWS 2003g).

There was also controversy over the role of the EA. Weston (2005), who headed the establishment of CMAs in South Africa, rehearsed the charge that 'the Environment Agency guys thought they were going to be mentoring us' and offering 'free consultancy' rather than partnering. But Weston himself (2005) rejected this charge:

> At this stage I don't see that. I'm looking for partnership on this, not to just be able to tap into their minds whenever we want . . . The Environment Agency have been looking at the European Framework Directive [which is designed to improve the health of the ecosystem, the quality of the water, and benefit the population and wildlife] . . . and the Environment Agency guys could see a lot of similarities in what Europe is aiming to do, to what we're doing. In a sense we are a couple of steps ahead, a year or two ahead maybe, so I think there was a realization that there could be some sort of learning from our experience, as well as us learning from what they're doing already – a two-way exchange . . . People are very wary of . . . the colonialists coming in again and saying you should do it this way . . . I think we have to . . . say this is a partnership, we want to be feeding into your processes as much as you are feeding into ours . . . whilst it's good to learn from other people, we do accept that we have to find our own way.

Clarke (2005) also saw a two-way process:

> There are an awful lot of similarities between the way the UK wants to do things and the way South Africa wants to do things, so the UK is also learning from South Africa on a lot of aspects . . . on things like the Water Framework Directive (WFD), the last thing we want to do as the UK is go and say, 'you should implement the WFD,' because it's appropriate for the EU but it's not appropriate for anywhere else . . . What we would say is that the tools that the EA are using to implement the directive could be helpful to the people in South Africa, and other countries as well . . . You cannot force your policies on to somebody else, it just doesn't work, but the way you implement them can certainly be used as examples of both good and bad ways to do things.

Seetal (2005) referred to a process of 'mutual sharing' of experiences:

> A lot of my colleagues were sceptical about what PAWS, or what water experts from Britain, could offer South Africa, largely from an experiential perspective. Without a doubt we could share expertise, but in terms of experience, the societies are pretty different . . . However, we were proved wrong in the sense that there are things that Britain can learn and things that we can learn as well, so there can be mutual sharing.

However, Quibell (2005) pointed out that there were limits to the extent to which the experiences of northern countries could grasp the significance of South Africa's apartheid past:

> Other countries may have similar social and economic divisions to South Africa, but what is different is that we come from an apartheid past. Our actions are therefore not only about addressing poverty and service provision but also about a political, social and moral obligation for redress. This, coupled with the fact that certain sectors of South Africa are well developed, means that we have to take from the 'haves' to give to the 'have-nots'. So I'm not sure international experience will really help us with that . . . there are very defined problems PAWS can assist with . . . but I'm not sure they can really make a significant policy/strategy contribution . . . they can deal with particular problems but they cannot deal with the context within which that problem occurs.

This was a point that Khambule (2005) readily acknowledged, saying that 'PAWS is very realistic in terms of saying, if I were in the UK I would do it this way, but because this is in South Africa I can only share my experiences, it may not be applicable in your circumstances.'

So, on the question of whether PAWS created real partnerships in South Africa, the verdict is mixed. On the one hand, as Oudyn (2004) argued, 'there is a real partnership between some municipalities in South Africa and the relevant people in PAWS,' which lent credence to interviewee 1's (2003) assertion that 'partnerships will be, if we have them, from the projects that we do.' Nel (2005) testified that in his own municipality:

> we have established a permanent direct link to our twins or partners in the UK . . . for us the permanent relationship has already been established . . . we are equal partners because they are learning a lot from us . . . it's aiding them to deal with similar types of issues in other areas so it's improving their position to give better support.

On the other hand, as Oudyn (2004) commented, even though there is some evidence of partnership between South African and UK stakeholders, 'on the whole . . . it's a British partnership working to do good in Africa.'

Uganda

In Uganda, as Interviewee 1 reported in November 2003, it was felt that PAWS could bring 'information, education, and facilitation around how to bring parties together' within the country. However, after undertaking a scoping visit in June 2003, PAWS realized that in Uganda those kinds of needs 'were already being satisfied' (Interviewee 1, 2003). As Causer (2005) noted, 'Uganda is actually a very popular destination for donors, and there are so many people trying to do useful things in Uganda.' Therefore, an NGO representative (Interviewee 3, 2003) said that it was decided that PAWS should do 'something quite light in Uganda because there is so much already happening there and [PAWS is] . . . yet another player with whom to co-ordinate.' PAWS decided to focus on what the water sector lacked, which was 'institutional frameworks and things like regulation' needed to underpin sector reforms (Interviewee 1, 2003).

In mid-late 2003, PAWS received a proposal from DFID and the Utility Reform Unit in the Ministry of Finance, Planning and Economic Development of the Ugandan Government, for the partnership to be involved in a regulatory exchange programme between the UK and Uganda, a proposal PAWS accepted. This resulted in a group of Ugandan stakeholders (mainly senior government officials and ministers) coming to the UK in November 2003 to receive 'a view of how you can make . . . [regulation] work within the resource base and realism of Uganda' (Interviewee 1, 2003). However:

> It was made clear that the intention of the programme was to give a clear and impartial view of the systems of regulation, which

> operate in England and Scotland, and the roles of various stakeholders. It was emphasized that the partnership was not endorsing or promoting a particular system, simply sharing its collective experience (PAWS 2004b, 30).

During this visit, the Ugandan delegates heard from various actors in the UK about regulation and 'how it affects different groups . . . the private sector . . . regulatory bodies like OFWAT . . . and NGOs' like Tearfund (Peacock 2003). As an NGO partner (Interviewee 3, 2003) commented, the event was 'small in a way, but the potential of it is to influence Ugandan regulatory systems and that could really benefit the poor if it's done in the right way.' The study visit was largely government-based on the Ugandan side, although it did include a representative from the Ugandan Water and Sanitation NGO Network (UWASNET) which is 'mandated to ensure [the] effective coordination, networking and collaboration [of] NGOs and CBOs in the water and sanitation sector in Uganda' (UWASNET 2006). Interviewee 1 (2003) remarked that 'if the processes in Uganda are working, i.e. that UWASNET is genuinely representative, then I think PAWS has chosen the right civil society partner.' As it happened, the UWASNET participant was a Ugandan WaterAid official because at the time they also held the UWASNET chair.

However, the bulk of the work which PAWS undertook in Uganda was subject to the same criticisms raised about much of PAWS's work in South Africa, in particular that it was more like free consultancy than partnership, and that it had not maximized the synergistic possibilities available through the multi-stakeholder partnership modality. Although UK-based NGOs, private sector and government actors were all involved in the Ugandan process, it was only in their giving expert advice to Ugandan delegates, not in their engaging as partners.

In addition to this work on regulation, in September 2004, PAWS was approached to support the Association of Private Water Operators [APWO] in Uganda, by helping to build their institutional capacity:

> This organisation was set up with World Bank support in 2003 and was established in order to provide effective management of water and sanitation services. In Uganda 50 out of 58 Urban Water Supply Authorities have established Water Supply Boards,

> and contracted the operations of their systems to private operators on cost recovery principles. The role of private water operators in provision of water supply and sanitation services in Uganda [including APWO] therefore seems to be widely accepted by stakeholders (PAWS 2005d).

Nonetheless, Turner (2005) maintained that PAWS would not work anywhere without the backing and support of the indigenous government in that country:

> we don't want to just go in at a low level. We can't work in a country unless we have political buy-in, so we don't want to engage with the Association of Private Water Operators, or go and do anything in one of the municipalities, unless it is under the umbrella of a MoU with somebody like the DWD [Directorate of Water Development].

While visiting Uganda to discuss support for the APWO, PAWS found that although donor organization and support in urban areas was excellent, there was a gap in support for water supply in rural areas. Turner (2005) reported that the impression given to the Secretariat from the annual donor sector review meeting in Uganda in 2003 was that 'Uganda is "full", they can't take any more help,' but he argued that 'we found when you get into the rural areas . . . that certainly isn't the case . . . in the urban areas they seem to be incredibly well organized, but there are still gaps that we think we could fill in the rural areas.' Therefore, following the Annual Sector Review in Uganda in September 2005, a MoU was signed between PAWS and the Ugandan DWD 'and their associated institutions to deliver capacity building support across several identified gaps within the existing framework' (PAWS 2006c). These projects 'include improved management of water operators, support to Water Authorities, technical support to Districts, [and] technical support on all aspects of water quality management' (PAWS 2006e).

As part of this programme, 'PAWS partners are providing training and accreditation support for water quality management in Uganda by ensuring that training staff are qualified and working to common technical standards' (PAWS 2006j, 11). Furthermore, PAWS also worked 'with the DWD and local NGO groups to identify how it . . . [could]

provide assistance to rural townships [that were] presently not receiving donor support,' which suggested that local civil society was playing a more conspicuous role in PAWS's work in Uganda than it was in South Africa (PAWS 2006c). Significantly, WaterAid and Tearfund were both active in Uganda, unlike in South Africa. In February 2006, a part-time country coordinator, Maimuna Nalubega, was appointed, and PAWS begun work supporting UWASNET to:

> improve its handling of the management skills of the founders of more than 120 member organizations. The members of the UWASNET member organizations have skills that can be utilised in the provision of water and sanitation in Uganda, but the potential is not being reached. PAWS has worked to address this and to provide ideas and suggestions on how to capitalise on the skills within the organisation and to establish mentoring support (PAWS 2006d).

However, this work faced political hurdles, as the PAWS May 2006 newsletter (PAWS 2006d) reported: 'UWASNET hopes to put . . . [the] suggestions [offered by PAWS] into practice to enhance the effectiveness of the organisation . . . [but] due to political pressures, this work may unfortunately be delayed.' Nevertheless, PAWS found some degree of synergy amongst its UK partners for its work in Uganda, and also involved local civil society actors, local private sector organizations, as well as government, although not necessarily in partnership with each other.

Nigeria

PAWS decided to engage with Nigeria through DFID, and, as John Barker (2004) of WWF explained, 'DFID directed us in quite a focused way to the four states in which they were working.' As a result, in September 2002, a scoping study team visited Enugu State and came back with proposals for four projects. Rather than concentrate on capacity building at the central and/or local government level, as was the case in early work in South Africa and Uganda, 'Part of the scoping study report [for Uganda] . . . focused on development of a project with a community in Enugu with an emphasis on technical and community sustainability' (PAWS 2004a). In November 2003, Peacock (2003) reported that 'we did talk at the village level and at the village water committee level' during the scoping visits to the country. An NGO member (Interviewee 1, 2003) observed that within Nigeria, 'WaterAid has been acting as a sort of bridge and advisor' to help 'bring the . . . Nigerian society element into the local equation.' This suggested that PAWS achieved greater synergy in Nigeria between its UK partners than it did in South Africa and in early work in Uganda. In Nigeria, alongside the UK private sector and government, because a strong role was found for at least one UK NGO (WaterAid), the role of Nigerian civil society in PAWS was promoted. However, Barker (2004) maintained in April 2004 that, 'Within Nigeria relationships are very poorly developed between government, civil society and the private sector . . . there's a failing government body that is . . . looking for a donor in some form or another and not really wanting or even having the capacity to develop a multi-stakeholder relationship.' By 2005, the initiative had not progressed beyond the stage of scoping visits within Nigeria. In fact, PAWS struggled to set up any real programme of work within the country throughout its first three years of operation, and this early work became, what Turner (2005) termed an 'abortive engagement'.

On the UK side, one factor explaining the failure of PAWS's early engagement with Nigeria was suggested by Turner (2005) – that:

> some of the private sector partners may have not had the right idea about what PAWS was about . . . [and] saw the process as being, [first] a feasibility study is done, and then a bit of work is let to a contractor. So there was a perception that there was paid work to be had somewhere down the line for somebody [when there was not].

On the Nigerian side, another factor explaining why this hiatus of action occurred was that Nigeria was 'going through its own assessment and developing its own poverty reduction strategy,' the National Economic Empowerment and Development Strategy (NEEDS) (Barker 2004). PAWS was reluctant to start any projects before this assessment process was complete, in case such work ran counter to the conclusions of the NEEDS, which was eventually launched in May 2004 (Peacock 2003). Interviewee 1 (2003) claimed that the lack of progress was also due to 'political concerns' and the weak governance structures at play within the country. Nigeria has a federal system, and in April and May 2003, elections took place for the national assembly, the president, and the governors of Nigeria's 36 states. It was widely reported that these elections were not "free and fair", with 'distribution of cash to voters . . . [and] electoral officials . . . the widespread lack of secret ballot . . . [and] endemic rigging attempts and sporadic violence' (Centre for Democracy and Development 2003). Thus, while 'Many hoped that the 2003 elections would represent the consolidation of civilian rule and democratic development . . . [they] fell short of these expectations and raise questions about the viability of the current democracy in Nigeria and also about the prospects of survival of the state itself' (Gramstad 2003, 21). The occurrence of these elections and the political turmoil surrounding them, led the PAWS process to stall, which showed the difficulty of working in partnership with government, let alone in partnership with government *and* other stakeholders, in countries where democracy is fragile. Ignoring government in such situations and working directly with other groups is tempting, but, as we have seen, the danger in that approach is of undermining attempts to build strong governing institutions within the country.

However, later activity in Nigeria was more promising, especially with regard to civil society involvement. In early 2005, an 'engagement with Benue State . . . [began] in partnership with WaterAid Nigeria' (PAWS 2005b, Annex C), 'building on pre-existing work that . . . [WaterAid] were doing' to support the water sector in small towns and peri-urban areas (Twelvetrees 2005). In February 2005, a PAWS team visited Nigeria, facilitated by WaterAid Nigeria, and 'met with key politicians and officials in Benue State, who were very keen to see PAWS participation in the state,' and also 'with a stakeholder steering group for the development of the state water sector,' to discuss 'the framework for PAWS involvement' (PAWS 2005d). Drawing on experience from South Africa regarding the capacity of African actors to receive PAWS support, PAWS's April 2005 newsletter (PAWS 2005d) stated that 'We were pleased to note that there is a structure set up with resources to support the delivery of services to the small towns which we believe has the capacity to receive PAWS input.' After the scoping visit to Benue State in February 2005, it was established that 'there was a need for the sort of capacity building . . . [PAWS] could give' (Turner 2005). The scoping team said that they would like to develop a MoU between Benue State, WaterAid Nigeria, and PAWS-UK, and this was subsequently completed.

In April 2005, the first working visit to Benue State took place, made up of the PAWS Chair, a Secretariat member, two private water operator partners, and a consultancy firm representative. Paul Turner (2005), the PAWS Manager, explained that:

> the State suggested . . . we pilot a model of community run water in three areas [the towns of Naka, Lessel, and Ughbokpo]. This wasn't something they had just thought up; it was part of federal policy guidelines and the basic principle is around the establishment of a Water Consumer Association, with the representative groups from the community, the transfer of the assets to that Water Consumer Association, and then that body really taking on the full responsibility for water and sanitation in that area . . . they wanted support in actually getting that established, initially in three pilot areas, because it hasn't been done anywhere in Benue State and we don't know whether the consumers are going to want to set it up, whether they will want to take ownership for it, what the issues are . . .

Turner (2005) reported that the type of assistance that PAWS offered in Benue State was 'project management . . . technical support, some training, [and] . . . technical assessment work.' In formal documentation, the areas of work listed for Nigeria were 'development of water services master plans, community based water initiatives, technical knowledge building and training' (PAWS 2006c) The working visit team looked at three very different towns and 'produced a report which gave a technical assessment . . . with issues, recommendations, and so on' of each, 'but the main body of the report was really an action plan for moving forward to do these pilots in the three towns' (Turner 2005).

Learning from experience in South Africa, it was decided to recruit a full-time local co-ordinator (or 'Country Programme Manager') for Nigeria, and Kayode Sanni was appointed in August 2005; in November 2006, Nyananso Gabriel Ekanem took over this role. Turner (2005) explained that, although 'there are central, federal policy units' in Nigeria, 'each state is pretty autonomous . . . and they generally make their own decisions on what they want to do.' As a result, although PAWS had the backing of the Nigerian central government, rather than have a National Steering Group in Nigeria (as was the case in South Africa), a 'Benue State Steering Group was formed.' Turner (2005) reported that the partnership resolved to do the same in the other Nigerian states that it engaged with. The Benue Steering Group was made up of the PAWS co-ordinator in Nigeria, a representative from the Benue NGO Network, WaterAid Nigeria, the State Department that was responsible for small towns within the Water Ministry, and another ministry representative who assisted with the Secretariat function for the group.

It seemed, as Turner (2005) declared, that in this work in Nigeria, PAWS was 'getting a much better engagement between those three sectors' – government, the private sector, and civil society – at least in terms of synergy between the UK partners, than it had in South Africa. Furthermore, although 'the private sector [in Nigeria hasn't been] engaged yet,' local civil society groupings have played a strong role in the initiative, and eventually Benue State might 'decide to use some small scale private sector operators in these towns' (Turner 2005). Turner (2005) put the very positive and conspicuous involvement of Nigerian civil society in this project down to the different approach employed here compared with PAWS's previous engagements. Complementing the link with state government was an engagement with local civil society facilitated by WaterAid. PAWS began this programme in partnership with WaterAid Nigeria, who 'know the country very well . . . [have] been [working in Nigeria] for at least fifteen years . . . and . . . [in] Benue State . . . [for] ten years . . . [and have] knowledge of the politics within the country, knowledge of the social aspects,' and also had links with the government and local NGOs (Turner 2005). Thus, by utilising the UK NGO partner WaterAid's capabilities and links in-country, PAWS was able to forge strong and stable relationships within Nigeria, which encompassed both the governmental sphere with 'political buy-in', and the NGO or civil society sphere, something which would have been difficult without WaterAid's assistance, as the initial 'abortive engagement' demonstrated (Turner 2005). Indeed, Turner (2005), speaking in June 2005, referred to WaterAid's knowledge on the ground, and ability to gauge the capacity of the State to accept PAWS's help, as vital:

> What has really come true in the last six months is the valuable knowledge that they [the UK NGO, WaterAid] have about

> in-country, and the contacts that they have, so I think we're starting to understand what we mean by the tri-sector partnership . . . WaterAid know what we offer and they know whether or not that State can accept our help, so we're not taking a risk . . . WaterAid wouldn't have taken us to Benue if they didn't know that Benue could actually take our help so that's what's really helped . . . [Benue] has been very easy to work with . . . they value our time . . . and they make sure they get the most out of it.

Aligning the partnership with ongoing work within Nigeria, and building a programme based on the foundations of one of its partner organization's on the ground experience and contacts, meant that PAWS was better equipped to deal with the 'particularly demanding circumstances of Nigeria,' and able to 'share the difficulties of working' in that country (PAWS 2004a). This is not to say that it was smooth sailing with the new Nigerian engagement, because the political situation in Nigeria continually impacted upon PAWS. For example, Turner (2005) explained that on one visit to Benue in 2005, 'we found that basically the cabinet had been dissolved and so the Water Commissioner we were working with was gone,' but he added that the appointment of a PAWS Nigerian co-ordinator would help the partnership to tackle such problems in the future.

WaterAid committed 5 million Naira (approx. £22,000) to the project; Benue State 'is putting in about 20–25 million Naira' (approx. £90,000 to £110,000); and PAWS-UK 'quantified the amount of technical and managerial resources' that they contributed as being approximately £20 million Naira (approx. £90,000) (Turner 2005). As Turner (2005) remarked, 'it's good from a partnership point of view because it's three parties that are contributing both people and money.' The money provided by Benue State, as well as being used to 'do extra [water] connections and things like that,' made it possible to employ three local NGOs with experience of working with communities to undertake some of the project work (Turner 2005). These NGOs 'actually do some of the community empowerment' and 'help with getting the principles established for the Water Consumer Associations' which the partnership assisted in setting up (Turner 2005). Turner (2005) reported that one of these NGOs was 'a strong environmental group, another focuses on women and children, and the other is an Anglican church group.'

PAWS also 'delivered training to the Small Towns Unit (a department within the Benue State Ministry, which is responsible for establishing the WCAs [Water Consumer Associations]) on aspects of cost recovery and water treatment technology' (PAWS 2006j, 8) to give further support to their development. The idea was that Water Consumer Associations would eventually be set up in 73 small towns in Benue State, which all suffered from 'poor infrastructure for managing the provision of water and sanitation services' (PAWS 2006j, 8). In 2007, PAWS 'engagement with Nigeria was to be extended to cover [a] further 3 states' (PAWS 2006e). Furthermore, the 2005–06 Annual Report (PAWS 2006j, 9) noted that 'ongoing support is being provided by the partnership by assisting in the small towns' application to the EU Water Facility to secure funds for a number of projects across the region.' Since April 2005, PAWS has worked 'in conjunction with the managers and operators' in Benue State 'to support the practical operation of water treatment works,' utilising 'On-site training . . . [and] ongoing distance support and mentoring' (PAWS 2006j, 9).

In April 2006, EA staff visited Anambra State in Nigeria to investigate 'the extreme and devastating effects of gully erosion,' and to 'explore whether integrated water resource management could help prevent further erosion and reduce the sedimentation of water bodies for water supply' (Twelvetrees 2005). Subsequent discussions between PAWS and the Nigerian Federal Environment Protection Agency on how PAWS could support the 'national gully erosion plan' (Twelvetrees 2005), showed further evidence of successful partnership work by PAWS in Nigeria.

Evaluation of Outcomes

As in the previous chapter, this case study chapter has concentrated on the processes of partnership and the ways of working of a TTP. Less attention has been devoted to the outcomes of the PAWS initiative and what it has achieved on the ground, because the first few years of PAWS's existence yielded few concrete outputs, so there is far less evidence available about what PAWS has accomplished than about how it has gone about doing its work. Nevertheless, there is *some* evidence of PAWS's achievements, and this is the area of discussion to which we now turn.

The sections on PAWS's work in Uganda and Nigeria are much shorter than the section on PAWS's work in South Africa. This is because, until recently, PAWS made limited progress in its engagements with Uganda and Nigeria. Nevertheless, the early efforts in Uganda and Nigeria provided important evidence about the processes of partnership and the lessons that PAWS has learned, and with the recent reinvigoration of work in these two countries, it is possible to evaluate PAWS's experiences in them. Unlike South Africa, we have no primary data from within Nigeria or Uganda on which to judge PAWS's success in them, and so we are reliant on the data that was available to us from interviews and documentation in the UK.

One of the difficulties in assessing PAWS's output is that it took the form of "soft" assistance, aimed at building capacity within its partner countries, rather than "hard" assistance, aimed at putting in "taps and toilets". The latter, although requiring some qualitative assessment, could be evaluated using quantitative criteria of numbers of people receiving water and/or sanitation services as a result of a project, but monitoring and evaluating capacity building is much more difficult. Turner (2005) noted that 'I don't think that we can ever put a definition or success criteria on PAWS that is ex-thousand people have now got access to water who didn't previously, it's not that sort of initiative.' Davitt (2004) explained that 'you can do all the capacity building you want and then it [might not] go anywhere . . . and there . . . [might be] no increase in service provision . . . My hope would be that as the capacity building swings into line and takes effect, the operational stuff behind it can get going.' The central problem is that it is difficult to definitively attribute improvements in service and delivery directly to PAWS. This is a problem common to TTPs because 'the majority of partnerships do not provide direct environmental benefits' (OECD 2006, 24). However, as Oudyn (2004) pointed out, just because it is harder to evaluate progress made in "soft" projects, does not mean that they do not have an impact: PAWS 'hasn't delivered mega litres of water [but that] doesn't mean that it hasn't done anything . . . it has had a real effect . . . it's been a step forward, if only a small one.'

Moreover, as Twelvetrees (2005) noted, PAWS still 'needs to be able to articulate achievements' – even if 'it is . . . more difficult to express qualitative rather than quantitative achievements . . . [PAWS] was set up to help meet the water and sanitation targets within the Millennium

Development Goals . . . [and] they have the responsibility to prove that PAWS is doing something towards that target.' A private sector participant (Interviewee 14, 2004) concurred, commenting that, 'we've got to deliver, we've got to have tangible results.' Nel (2005) observed that although it was very difficult to monitor PAWS, 'because it's actually setting up systems and institutional arrangements that will improve service delivery,' progress is 'definitely measurable, you just have to be innovative.' In the end, as Khambule (2005) commented, 'it's not so much about how many visits they've made . . . how many documents they've reviewed, and how many municipalities they are supporting . . . how much money has been spent . . . it's about the impact on the ground,' and there has got be some means of monitoring that.

Turner (2005) revealed that after he joined the Secretariat, a business plan was developed which aimed to address this need for output-oriented monitoring by identifying 'specific outputs . . . with success criteria.' Although Turner (2005) conceded that 'some of those success criteria are a bit subjective,' he maintained that 'at least we have a set of measures' now. In the minutes of the PAWS Steering Group meeting in May 2005 (PAWS 2005b), a report was made on PAWS's performance, using the success criteria established in the 2005/06 Business Plan. Here eight success criteria were listed, to which were attributed various weightings. On the first criterion – 'Delivery of tangible projects or sub projects' – it was reported that while formal project procedures had been established in a Partnership Handbook, and that projects were well defined in Nigeria and Uganda, 'Projects in SA [South Africa] . . . [had] not materialised as expected' (PAWS 2005b, Annex B). This referred to the problems encountered in establishing programmes with both the initial four pilot municipalities, and the additional five. On the second criterion – 'Imparting knowledge which is not otherwise available' – the partnership judged that it was fairly successful, giving examples of the EA work in South Africa, water treatment work in Benue State, and the work Kelda were doing with KwaZulu-Natal (although this continued to be at a low level) (PAWS 2005b, Annex B).

On the third criterion – 'Delivery of tangible projects and support as a tri-sector partnership' (PAWS 2005b, Annex B) – it was noted that PAWS had not succeeded in establishing tri-sector partnering in South Africa, but that 'Good progress has been made with Nigeria which has the potential for being the first real test of the tri-sector approach'

(PAWS 2005b, Annex B). On the fourth criterion – the establishment of 'long term twinning to give confidence' (PAWS 2005b, Annex B) – disappointingly, the only two examples of this were with the EA and Kelda in South Africa, which were working in only two municipalities (PAWS 2005b, Annex B). On the fifth criterion – 'Measurable progress in "capacity building" within the recipient Municipalities and making a difference in the day to day management of water and sanitation' – was assessed as 'very disappointing' in South Africa, and the need for a formal review was asserted (PAWS 2005b, Annex B). On Nigeria and Uganda, the report stated that a programme was established in Nigeria and there was potential in Uganda; in other words, these programmes had not progressed far enough to be assessed in any greater depth (PAWS 2005b, Annex B). On the sixth criterion – 'Active participation of the majority of UK partners on projects within 12 months' – there was judged to be some success in 'New partners engagement with SWW [South West Water], Atkins, Severn Trent in Nigeria.' Furthermore, some new partners had joined, including 'OFWAT, Cranfield Uni, [and] Develop' (PAWS 2005b, Annex B). On the seventh criterion – 'Formal cooperation agreements for each country/municipality' – it was recorded that there was a MoU signed with Benue State in Nigeria, and another planned with Uganda, but 'Progress with SA [South African] municipalities [was] very slow' (PAWS 2005b, Annex B). On the eighth and final criterion – 'Impact on MDG targets' – which was assigned considerable weighting in terms of importance, the performance report conceded that 'No measurable progress' had been made (PAWS 2005b, Annex B).

Overall, therefore, the evaluation in May 2005 of PAWS's work on these criteria judged it to be only moderately successful, with a disappointing performance in South Africa, and only hopes of future success in new engagements with Uganda and Nigeria. However, the PAWS *Annual Report 2005–06* (PAWS 2006j, 9) noted some promising signs of progress in Nigeria, stating that: 'The result so far has been improvement in water quality and reliability to around 10,000 people with the intention to roll out across the entire state, which has the potential to provide safer water and sanitation to thousands more people.'

In addition to this formal assessment by the criteria established by the partnership itself, a second way of evaluating PAWS's success is to establish how African partners felt about the work done by PAWS. The assumption here is that as an initiative that offered no direct funding,

African participants would be unlikely to enthuse about PAWS unless they felt a real benefit from it. There is more evidence of consumer opinion in South Africa than in Uganda and Nigeria, because PAWS's work in Uganda and Nigeria was more recent. Much of that opinion in South Africa was positive. For example, Twelvetrees (2005) argued that 'when the South Africans came over from the different municipalities they seemed very happy with the relationship and their training, so that shows some success.' Also, Nel (2005), who worked for Zululand District Municipality, one of PAWS's pilot project areas, stated that 'The timing and support that's given is very good . . . PAWS actually suited our needs, from very little time spent with the guys there could be such a lot of benefit . . . there is major buy-in and massive support from our municipality' for PAWS. Similarly, Japie Visser of Nkomazi Municipality stated that 'a lot of progress has been made due to inputs given by the PAWS delegation in the past,' while George Mohlakoana, of the South African local municipality of Mangaung, who attended the study visit to the UK, asserted that 'we really learnt a lot' (PAWS 2005e, 14–15). Furthermore, Eddie Reynolds of uThungulu Municipality declared that 'We have benefited from the PAWS support and our request is that our mentors from the UK spend more time with us. Please continue with this support; we have learned from the PAWS initiative' (PAWS 2005e, 14). Furthermore, the fact that work expanded from four municipalities to nine, suggested that South Africans, at both the local and national levels, saw PAWS as a worthwhile initiative. Legoabe (2005) recalled that 'I've heard from the senior people in the DPLG that they are quite satisfied. I am sure that if that wasn't the case then nobody would be prepared to say, "let's recommend more municipalities," we wouldn't be wasting our time if there had been no impact' (Legoabe 2005).

A third way of assessing PAWS's outcomes is to examine its impact on its host country. In other words, beyond outcomes within Africa, it is also possible to judge whether PAWS has had an impact within the UK. As with the EUWI, an NGO participant (Interviewee 1, 2003) reported that one outcome of the PAWS initiative related to the improved understanding that was gained between the several stakeholders involved: 'Partnership is about negotiation around understanding each other more . . . not necessarily agreeing but getting a higher level of understanding . . . for those that have stuck the course, that's what's been happening.' Likewise, another NGO representative (Interviewee 3, 2003) claimed their organization would not have such 'good relationships with the partners [from] . . . business and . . . government . . . if it wasn't for PAWS,' and added that it was 'really useful to get that understanding of the bigger picture and understand where other people are coming from.' This means that PAWS was fulfilling one of its objectives – to encourage best practice in multi-stakeholder partnering. Two private sector players also reported the beneficial impact PAWS had had on their organizations: Davitt noted that for Kelda Water, the work they undertook in South Africa provided their staff with 'new insights into familiar challenges faced in very different circumstances to those in the UK;' while Turner commented that PAWS 'is . . . providing a valuable opportunity for . . . Northumbrian Water [staff] to enhance their personal development using their skills in a completely different environment. This benefits both the individual and the company' (PAWS 2005e, 17).

Shift in Direction

PAWS decided on a shift in its country selection policy away from countries like South Africa, which were on track to achieve their MDGs in water and sanitation services, towards countries like Ethiopia, which were far from on track. This shift was precipitated by PAWS's frustration at its lack of progress in South Africa. In May 2005, for example, PAWS-UK expressed serious misgivings about the water and sanitation supply work they had been engaged with in South Africa, and at a Steering Group meeting (PAWS 2005b, Annex E) it was noted that:

> A significant amount of effort has been expended in Municipalities over the last three years . . . This has resulted in some capacity building and knowledge sharing but there has been no long term engagement on proposed work programmes. The overall conclusion is that there is a lack of political will and in some cases lack of technical/managerial capacity within the Municipalities. As a result of this PAWS may never make any progress until more fundamental institutional reform takes place. In summary the situation is as follows:
>
> • The only successful activity in the last 6 months has been the EA working visit . . .
>
> • Both old and new municipalities were invited to the Steering

> Group in May to present their future needs. Most did not attend and only Zululand prepared a bid.

As a result of this frustration, from 2006 there was 'a scaling down of activity in South Africa due to declining demand and a shift of focus from in country work to remote support' (PAWS 2006d). Although PAWS did not abandon South Africa altogether, it concentrated much more on Uganda, Nigeria, and the new partner country of Ethiopia. This shift was also partly due to PAWS's perception that it had had a positive impact in Nigeria, especially in its engagement with Benue State. The fact that work was expanded to include three more Nigerian states was an indication that within Nigeria there was also a perception that PAWS had had a beneficial impact. Similarly, although the impact of PAWS in Uganda is difficult to judge, the signs are positive, not least because it has been more inclusive of non-governmental stakeholders within the country than was the case in South Africa, and because it has utilized synergies between its UK partners to a greater extent.

However, the main reason for the shift in PAWS's focus was because South Africa was less in need of help than were other sub-Saharan African countries. PAWS was now trying to engage only with those countries that were not on track to meet the water and sanitation MDGs. The partnership had developed country selection criteria, and Turner (2005) explained that since PAWS began to apply these criteria, 'it's obvious where we should be going because we're looking at the countries that have got the biggest gap to fill in terms of achieving the Millennium Development Goals, but at the same time are politically stable enough for us to work.' Clearly, South Africa would not have fulfilled the criterion of being in the most need. Turner (2005) reported that, at the annual meeting of the PAWS Forum in 2005, 'members agreed that we should go for another country, so that's why we're going to start an engagement in Ethiopia.' An initial scoping visit was undertaken by two members of the Secretariat, a law firm representative, and the PAWS Chair, in September 2005 'to meet with a variety of organizations within Ethiopia to ascertain the level of support required' (PAWS 2005c), and a MoU was signed in early 2006 with the Minister of Water Resources (PAWS 2006c). The areas of support were to 'include work at a federal level with the Water Ministries and local level support at regional districts . . . [as well as support on] regulation of the Addis Ababa public taps' (PAWS 2006e).

Turner (2005) noted that, as in Nigeria and to a lesser degree Uganda, PAWS was 'talking to DFID, WaterAid and Tearfund,' because they all had links within Ethiopia. In fact, the initial scoping visit was organized by WaterAid in Ethiopia (PAWS 2005c), to ensure that PAWS did not duplicate existing work programmes, and to establish strong multi-sector relationships within the country, find suitable projects, and understand better the environment within which they would be working. Furthermore, Roe claimed in December 2005 that PAWS would be 'working with NGOs [presumably Ethiopian] to identify projects for implementation over the next few months,' which was another positive sign that PAWS's approach to in-country work had improved since it first began its engagements with South Africa, Uganda, and Nigeria. In the *PAWS Newsletter July/August 2006* (PAWS 2006i), the PAWS Chair reported that 'a country manager to oversee implementation of . . . [the] work plan and to co-ordinate PAWS's work in Ethiopia' was to be appointed, and Abate Sileshi assumed that role in September 2006. This newsletter (PAWS 2006i) explained that in Ethiopia, PAWS had a '12 month programme of work . . . with a balance of projects at both a federal and regional level . . . [including] assistance . . . [on]:

- Sector monitoring and performance
- Capacity building with sector coordination teams
- Water treatment, networks and leakage support
- Benchmarking with Government enterprises.'

The *Annual Report 2005–06* (PAWS 2006j, 12) included a reference to PAWS's work programme for Ethiopia, stating that:

> The delivery of the programme will begin with a technical assessment of the Awassa Water Treatment Works, to develop a strategy report. This report will provide guidance and advice on how to secure funding to enable the . . . plant to continue providing a water service to a population of 104,000. This will enable the town water utility to make a structured proposal for expansion of the facilities which serve a population of around 250,000 people in the area.

Conclusion

In this chapter on our second case study – PAWS – we have described its origin in the UK with twin aims of process (to develop the partnership model) and outcomes (to improve water and sanitation provision by building capacity on the ground). With regard to process, the UK government anticipated that PAWS would evolve into a self-sufficient organization, but it remained dependent on DEFRA for its funding and secretarial support until 2007, when the funding role shifted to DFID and the Secretariat moved to Loughborough University. This meant that PAWS was perceived by many as a northern-led, and top-down (i.e. government-led), initiative for a large part of its history. In its work in South Africa, PAWS tried to counter that perception by involving southern stakeholders, but although it succeeded in engaging with indigenous state bodies (both nationally and locally), it failed to embrace the local private sector and local NGOs, or find synergy between its northern partners.

In Uganda, because urban areas were relatively well-provided with water and sanitation services, PAWS concentrated its efforts on rural townships, where it focused on training and capacity building. For instance, PAWS supported the work of an indigenous NGO network made up of 120 organizations. However, in Uganda, PAWS was criticized for being less of a partnership, than a dispenser of free consultancy. In Nigeria, PAWS struggled for three years to find an effective role. This was partly because northern private sector partners mistakenly thought that contracts were in the offing, and partly due to political concerns, including the fact that the Nigerian government was conducting a major poverty reduction strategy review. Later, however, PAWS made more headway, largely through enhancing the role of the northern NGO partner WaterAid in its operations, which in turn led to the greater involvement of indigenous civil society in-country.

With regard to outcomes, they are difficult to estimate, because much of PAWS's work has centred on capacity building, which is not easily reducible to quantitative assessment. On eight criteria self-assessed by PAWS in 2005, its performance in South Africa was disappointing. However, since 2005, PAWS has been more successful in meeting the criteria, especially in Nigeria. Therefore, PAWS also showed itself capable of learning from its experiences – not least in selecting more carefully the African countries where it could make a difference.

In the next chapter, we turn to our third case study – the West Africa Water Initiative – where we find less attention paid to process and more emphasis on outcomes, than in either of the first two cases.

5

THE WEST AFRICA WATER INITIATIVE (WAWI)

Introduction

This chapter, which focuses on the West Africa Water Initiative (WAWI), should be viewed, less as a complete case study (like the chapters on PAWS and EUWI), and more as an additional illustrative example of a TTP for water and sanitation in Africa. The data on WAWI was collected from interviews with representatives in the northern hemisphere head-offices of partner organizations; from attendance at a partnership side event at the 12th meeting of the CSD in New York in 2004; and from documents relating to the partnership. We begin our analysis by first, examining the origins of the initiative, its aims and objectives, and its relation to the TTP framework. Second, we discuss the membership and organizational structure of the initiative, and third, we consider how the partnership has operated at the implementation level. Finally, we conclude with a review of WAWI's achievements and an evaluation of the progress the partnership has made.

The Initiative and its Aims

A few days before the WSSD, in August 2002, Steven M. Hilton, president of the Conrad N. Hilton Foundation (CNHF) (a philanthropic organization established in 1944 by the hotel entrepreneur, with assets of around $2.5 billion), and Andrew S. Natsios, Administrator of the United States Agency for International Development (USAID), announced the launch of a new 'public private partnership to provide potable water and sanitation to rural villages in Ghana, Mali and Niger,

West Africa' (CNHF 2002). The initiative was to begin in October 2002 (World Vision 2002), and was 'Inspired by the vision of the Conrad N. Hilton Foundation' (Water for the Poor 2002c). It was built on the foundations of a 12-year project in Ghana – the Ghana Rural Water Project – which began in 1990 and was administered and managed by World Vision, with funding from USAID and the CNHF (CNHF 2002). In 2000, the CNHF explored the possibility of expanding their 'commitment to support potable water projects . . . in West Africa,' and so began talking to other potential funding agencies, which was how their relationship with USAID started (Interviewee 15, 2004). They issued a call for proposals and identified organizations to implement the new programme in West Africa, and some of the initial partners introduced new members: for example, USAID introduced the World Chlorine Council (WCC), while the Desert Research Institute (DRI) introduced Winrock (Interviewee 15, 2004; Interviewee 25, 2004).

The initiation of WAWI meant that the previous programme in Ghana could be expanded, and also that assistance could be extended to Mali and Niger. Mali and Niger had been identified as two of the countries most in need of assistance, and also where partners had the capacity to help (Interviewee 15, 2004): 'Mali and Niger are among the bottom five to ten poorest countries in the world' and were chosen 'based on need,' but also based on their 'proximity to Ghana' because 'the staff capacity . . . was already built in Ghana' and could thus be more easily 'exported to support these two' neighbouring countries (Interviewee 28, 2005). WAWI brought an additional peri-urban focus to the previous programme focus on rural areas.

Work on implementing WAWI's programme started in Mali at the end of 2002, but did not begin in Ghana and Niger until 2003 (USAID 2006). Indeed, the WAWI Coordinator, Nerquaye-Tetteh (2005), reported that WAWI partners were not 'fully deployed in the field . . . [until] October of 2003.' The majority of WAWI's work was focused in the same areas as World Vision's previously established Area Development Programmes (ADPs). In Ghana, the initiative targeted eight districts in the northern region; in Mali, nine districts in the Segou Region and two districts in the Mopti Region; and in Niger, two districts in Zinder Department and three in Maradi Department (CNHF 2004b; 2004c; 2004d).

Although WAWI was created 'as part of the global movement towards partnership' which was 'formally endorsed' at the WSSD, it was not initially registered with the CSD Secretariat as a TTP (USAID Water Team 2004b). Instead, its official relationship to the WSSD came under the banner of the US government's 'Water for the Poor Signature Initiative', 'which proposed an investment of up to $940 million' over the three years following the WSSD, 'to work with other governmental and non-governmental partners' on water-related projects (Murray 2002). Water for the Poor was 'a new platform for partnerships' (Water for the Poor 2002a), registered with the CSD Secretariat at the WSSD under the heading of 'activities to initiate partnerships', 'which have the potential to lead to the formation of new partnership initiatives for sustainable development' (UNDSD 2005).

From the very beginning, there was controversy between WAWI members over the initiative's relationship to the MDGs and the TTP framework. The US government listed WAWI under their own efforts to help achieve the WSSD's goals, making it clear that for the US government, and therefore USAID, it was important to highlight WAWI's role in the implementation of international targets, indicating their leaning towards having WAWI registered as a TTP. Another member (Interviewee 24, 2004) argued that having the initiative registered as a TTP would be valuable in drawing 'more attention to the region and the actual project, and possibly funding needs.' However, this interviewee (24, 2004) conceded that, although 'we haven't got consensus' on this issue, it is not 'because we don't want it to be recognized,' it is just that 'a lot of the partners, [such as] Winrock, [and the] Hilton Foundation . . . [are] more interested in getting the work done . . . getting the project started and working through the implementation phases.' Another member (Interviewee 15, 2004) also interpreted the issue as a top-down/bottom-up clash:

> it's great if it [WAWI] links to the broader initiatives and helps to meet the goals set at international conferences, but our focus has always been on the work in the field . . . Rather than a top-down approach with groups coming together at the international level and saying this is what we're going to do, and then letting it filter down to be implemented in West Africa . . . it should be driven the other way around.

With a limited staff-base and budget, WAWI's focus has been on ensuring that their 'money is having an impact,' rather than linking their efforts to international frameworks and goals (Interviewee 15, 2004). As another interviewee (24, 2004) explained, 'I think we are all behind it [getting registered] . . . it's just a matter of who wants to take the time away . . . [from other activities] to actually get it registered.' WAWI did eventually get registered as a TTP, although it is not clear when this took place.

The Chlorine Chemistry Council (2004) reported that the logic behind WAWI was that 'WAWI partners combine resources and skills to leverage funding from public and private sources, gain cost efficiencies, increase advocacy power with government policy makers, and develop more innovative and effective models of action.' The USAID Water Team (2004b) maintained that such 'Collaboration with other organizations creates programmatic synergy and accesses the complementary strengths of a number of affiliated partners.' Although, for the CNHF, the 'core emphasis . . . remains the link between water and human health,' WAWI recognized 'the need for attention to [be given to] a broader water management context' (Water for the Poor 2002b), and 'focuses on improving public health, providing increased water supply and sanitation services, and promoting sustainable, integrated water resources management in poor communities' (Chlorine Chemistry Council 2004). The USAID Water Team (2004b) stated that WAWI's wide-ranging workload included 'well-drilling and rehabilitation, hand and solar pump installation, alternative water source development, construction of latrines, household and school based sanitation and hygiene education, community mobilization and governance, hydrogeological analysis, capacity building, and policy development.' To this list, the CNHF (2004a) added that WAWI was also involved in the 'improvement of existing water and sanitation sources . . . provision of eye health medical supplies and training of health care personnel, field test of small-scale farm micro-irrigation systems, resource conservation, capacity building/training at community, government and local artisan levels . . . [and] action research' (CNHF 2004a).

WAWI was set up with a life expectancy of six to seven years, with the following stated aims (CNHF 2004a):

1) To increase the level of access to sustainable, safe water and environmental sanitation services in rural and peri-urban communities
2) To decrease the prevalence of waterborne diseases including blinding trachoma, guinea worm and diarrhoea

3) To ensure ecologically and financially sustainable management of water quality and quantity; and,
4) To foster a new model of partnership and institutional synergy to ensure technical excellence, programmatic innovation, and long-term financial, social and environmental sustainability in water management that may be replicable elsewhere.

These overall objectives were divided into specific expected outcomes, as the table shown on the next page (taken from the WAWI website) shows (Water for the Poor 2005a):

In 2004 (CNHF 2004a), WAWI set itself the following targets to be reached by 2008:

> provide a minimum of 825 wet wells equipped with hand pumps, reaching more than half a million people; 100 alternative water systems; 9,000 household and public latrines; and construct maintenance and repair technical resource centres, in addition to training thousands of adults, children and teachers in improved hygiene and sanitation practices, and increasing the skills of community members on development techniques together with water and sanitation practices.

In 2006 (WAWI 2006), these targets were raised to:

- 865 boreholes
- 420 rehabilitated boreholes
- 150 alternative water sources
- 110,000 [sic] latrines
- small-scale irrigation, hygiene promotion, IWRM, gender mainstreaming, advocacy, etc. interventions
- Total population receiving benefits: 650,000

The fact that all of WAWI's original targets were increased (although the figure for latrines is almost certainly a misprint), suggested that it was delivering its objectives. It also indicated that WAWI was attracting additional funding and partners. Starting in 2002, with $40.7 million and ten partners, by 2006, WAWI had $45 million and thirteen partners, or associated bodies (as explained below) (WAWI 2006).

Overall Goal: To improve the health and well-being of families and communities in Ghana, Mali and Niger			
OBJECTIVE 1: Safe Water and Sanitation	**OBJECTIVE 2: Disease Reduction**	**OBJECTIVE 3: Water Management**	**OBJECTIVE 4: Effective Partnership**
Outcome 1: Rural households have access to adequate all year round supply of water through increase in numbers of sustainable potable water sources.	**Outcome 1:** Increased community awareness and understanding of prevention of trachoma, guinea worm and diarrheal diseases.	**Outcome 1:** Communities (both genders) mobilized, organized and empowered to own and manage water facilities for sustainability.	**Outcome 1:** WAWI HQ and Country teams operational with shared visions well committed to the program.
Outcome 2: Hygiene and sanitation facilities in place and in use.	**Outcome 2:** Communities practicing appropriate behaviors for the prevention of trachoma, guinea worm and diarrheal diseases at the household and individual levels.	**Outcome 2:** Enabling environment created. **Outcome 3:** Sound environmental management practiced.	**Outcome 2:** Partnership defines WAWI strengths (including activities, tools, approaches) to be shared and harmonized.
Outcome 3: Residents of low-income urban settlements have access to water and adequate sanitation services.	**Outcome 3:** Increased awareness by teachers and school children and understanding of prevention of trachoma, guinea worm and diarrheal diseases.	**Outcome 4:** Livelihood and income generation promoted.	**Outcome 3:** Effective WAWI Project managed and compliant with donors, governments and community standards and procedures.
Outcome 4: Expanded water availability for agricultural purposes (drip irrigation and livestock watering) at selected villages.	**Outcome 4:** School children (boys and girls) and teachers practicing appropriate health, hygiene and sanitation behavior.	**Outcome 5:** Research capacities developed and research findings being utilized.	**Outcome 4:** Learning outcomes in terms of lessons learned.
Outcome 5: Increased efficiency in the development of WATSAN [water and sanitation] services.	**Outcome 5:** Integration of health and hygiene promotion into school curriculum.	**Outcome 6:** All local/community partners work collaboratively with communities for sustainability.	**Outcome 5:** Enhanced and unified institutional capacity for government and communities.

Membership and Organization

WAWI comprises ten core partners and three additional organizations. The ten core partners are: on the governmental side, one governmental bilateral donor organization (USAID); from civil society, three charitable/not-for-profit NGOs (World Vision, WaterAid, and Winrock International), one public international organization (UNICEF), two academic/consultancy research institutions (the DRI, and Cornell University's International Institute for Food, Agriculture and Development), and two grant-making organizations (the CNHF and the Lions Club International Foundation); and from the private sector, one global network of national and regional trade associations (the WCC). The three additional organizations are two NGOs (Helen Keller Worldwide (HKW), and the International Trachoma Initiative (ITI)), and a public policy organization (The Carter Center) (Interviewee 15, 2004; Interviewee 28, 2005; Interviewee 29, 2005; USAID 2002a; USAID 2004a; USAID Water Team 2004a; USAID Water Team 2004b; Water for the Poor 2002b; Water for the Poor 2002c). These three organizations are sometimes referred to as partners, but at other times as agencies acting in an advisory capacity to the initiative, because they do not receive grants for WAWI work, and do not act as funders themselves in the initiative. Most of the individual members' headquarter offices are situated in the USA, with one partner organization based in the UK.

Each partner has a specific role with defined responsibilities, and members can be divided broadly into two categories – funders and implementers – although some partners, such as World Vision, fit into both categories. The partnership has a MoU which is not legally binding, though 'the individual contracts that the partners have with the donor agencies' such as USAID and the CNHF, are 'legally binding documents' (Interviewee 28, 2005). The initiative does not actively seek new partners who solely want to undertake implementation work, but it does court new funding partners (Interviewee 15, 2004). The CNHF, as a private charitable foundation, 'is the primary external donor' for WAWI with 'an important coordination and oversight role' (Water for the Poor 2002b). The CNHF has given out grants on the basis that these are 'match-funded' by other partners (CNHF 2004a). WAWI began life with pledged contributions of nearly $18 million from the CNHF, over $16 million from World Vision, $4.4 million from USAID, and smaller contributions from other partners, bringing the total initial funding to $40.7 million (CNHF 2002). World Vision serves as the 'lead agency for the implementation of the program' (World Vision 2002), and:

> At the request of the Hilton Foundation, World Vision has created a regional water and sanitation technical team responsible for the installation and management of all Hilton-funded hardware, or borehole interventions . . . [Furthermore], World Vision will capitalise on the strength of its Area Development Programmes to facilitate and train local communities in latrine and laundry pad construction, pump maintenance and repair, and women's literacy (CNHF 2003).

In 2003, the roles of the other partners were stated as follows (CNHF 2003):

> In close partnership with World Vision staff in Ghana, Mali and Niger, the Desert Research Institute will offer technical assistance and training to national staff in borehole siting, hydrogeological mapping, water quality analysis and conservation of water resources. In northern Ghana, Winrock, in collaboration with Desert Research Institute, will field test and introduce a micro-irrigation pilot project among small-scale farmers for increased food security and income generation . . . Lions Clubs will focus on improved health and hygiene practices in Mali and Niger, in an attempt to reduce prevalence of unnecessary blindness; UNICEF will support capacity building of rural water supply programmes at the national and sub-national levels. With Hilton funding, UNICEF will also introduce school sanitation and hygiene education programmes in rural primary schools in Ghana, Mali and Niger. In both Ghana and Mali, WaterAid will bring together and supervise local latrine artisans and hygiene promotion activities. In addition, WaterAid will provide community standpipes, latrines and domestic waste pits to improve access to water for the urban poor in Bamako, Mali. Cornell University's International Institute for Food, Agriculture and Development will continue its Hilton-funded collaboration with World Vision Ghana through its Natural Resource Management and Sustainable Agriculture

> Partnership. These projects focus on resource conservation around wet boreholes, experimentation with a system of rice intensification, community-based land use planning and management, household and farm-record keeping and farmer-centred research and extension. Building on the institutional strengths and relationships developed in its guinea worm, river blindness and trachoma programmes in Ghana, Mali and Niger, the Carter Center will help WAWI partners make contact with the public and private health care sectors in each country. It will also provide help to identify communities with the highest incidence of preventable diseases . . . Helen Keller Worldwide is making use of Hilton Foundation support for trachoma control in 12 countries, including Mali and Niger. WAWI's partners will benefit from knowledge gained through HKW's comprehensive trachoma control approach in both countries, particularly those strategies adopted by national public health systems.

USAID provided grants (which did not require match-funding) to six partners (World Vision, the DRI, Cornell, WaterAid, Winrock and UNICEF), which were administered by Associates in Rural Development (ARD) – 'a firm that does a lot of business with USAID in a number of areas' (Interviewee 26, 2004). USAID also contracted ARD 'to provide three years of . . . technical assistance' (Interviewee 26, 2004), concentrating on strengthening 'the integrated water resources management orientation of the initiative through support to livelihoods and income generation, policy and enabling environment, gender mainstreaming, and hydrologic information management in both rural and peri-urban settings' (CNHF 2002). The 'World Chlorine Council – in conjunction with the Vinyl Institute, donated PVC pipe wells, chlorine for water disinfection and education materials on sanitation and hygiene' (CNHF 2002), and the ITI worked in all three countries, focusing on the construction of rural hygiene and sanitation facilities (Interviewee 15, 2004; Water for the Poor 2005b; 2005c; 2005d).

WAWI has multiple levels of governance, which one partner described as 'pretty loose' (Interviewee 25, 2004). The Headquarters Steering Committee is made up of representatives from all thirteen WAWI members, 'and provides overall policy guidance and direction for the partnership' (Nerquaye-Tetteh 2005). One participant (Interviewee 26,

2004) said that 'their responsibility is to provide backstopping decisions that allow the people in the field to carry out their work . . . mostly it's related to funding.' However, the headquarters level grouping also provides 'specialized help in terms of high-level or high-tech issues that the field will need' (Interviewee 28, 2005). For example, there is a Water Quality Working Group, headed by Braimah Apambire of World Vision, established 'to help the field deal with certain water quality protocols and issues' (Interviewee 28, 2005). Thus, as one Headquarters Steering Committee representative (Interviewee 28, 2005) explained, 'We try to empower the field to do everything but if there are some specialized things that we need to take a lead in, and then get the field on board, then we'll try to do that.' There is also a Strategic Planning Working Group, a Monitoring and Evaluation Working Group and a Fundraising Working Group (Interviewee 29, 2005). The Headquarters Steering Committee meet roughly twice a year and its meetings are professionally facilitated, although the agenda is set by the Secretariat (Interviewee 25, 2004; Interviewee 26, 2004). One interviewee (15, 2004) explained that decisions within the initiative are rarely achieved through voting, but through 'operating by consensus.'

According to one WAWI member (Interviewee 17, 2004), WAWI has no 'central controlling entity'. Doyle and Corliss (2006, 1) explained that 'An explicit decision was made . . . that WAWI is defined as a function of the identity, strength and presence of its individual partner organizations, and that a significant amount of funds would not be diverted into building a new organization for the partnership itself.' However, the WAWI Secretariat still holds a pivotal role. It is supposed to be an independent body but some partners view it as run by World Vision (Interviewee 23, 2005; Interviewee 27, 2004; Interviewee 28, 2005; Interviewee 29, 2005). For example, one headquarters representative (15, 2004) stated that 'the Hilton Foundation designate World Vision as the provider of the Secretariat . . . World Vision supports most . . . of the Secretariat's financing' (although not all), and provides it with office space. Furthermore, the Secretariat is headed by the WAWI Coordinator, Bismark Nerquaye-Tetteh, who was previously the director of World Vision Ghana. One WAWI partner (Interviewee 23, 2005) claimed that World Vision were not 'very happy' about the fact that 'they had to take somebody out of their programme' to form the Secretariat.

The Secretariat is based in Accra, Ghana and a 'small group [of] around four or five people' work alongside Nerquaye-Tetteh (Interviewee 26, 2004). Nerquaye-Tetteh attends meetings both within Africa and at the headquarters level, and the Secretariat is responsible for overall coordination of the initiative. Nerquaye-Tetteh is the lynch-pin of the initiative, described by one member (Interviewee 24, 2004) as 'like the Executive Director overseeing the whole project . . . communicating the needs of the three countries back to the headquarters partners, and . . . communicating what we [headquarters partners] want to the people working on the ground.' Another participant (Interviewee 26, 2004) stated that 'Bismark's job is to co-ordinate the activities, and, based on the progress reports and field visits, assess how things are going . . . [He has to] try to facilitate collaboration, and, based on work plans, make sure that everybody is coming together to fulfil their promises and responsibilities.'

There are also World Vision-led WAWI Country Teams and National Steering Committees in each of the focal countries, made up of representatives from all of the implementing partners within each specific country (Nerquaye-Tetteh 2005). These in-country groupings are intended to act 'in coordination, and in full consultation with government officials in each country' (WAWI 2006). National Country Teams meet approximately 'every couple of months' (Interviewee 26, 2004), to discuss 'operations and activities' (Nerquaye-Tetteh 2005). On an annual basis, WAWI partners from the three participating countries meet 'to share progress results and discuss future strategies' (Nerquaye-Tetteh 2005). Evidently, 'partnering is a lot stronger within each country than it is cross-country' (Interviewee 25, 2004). Some implementing partners, such as World Vision, have many on the ground field staff within the focal countries. Others, such as the DRI and Cornell University, send people in, as and when required.

Alongside partners who are based in the northern hemisphere, 'The Initiative will collaborate closely with [southern] governments and other local actors who will be participants at all stages of activity design and implementation to maximise the impact of water-related interventions by public and private actors alike' (Water for the Poor 2002c). Indeed, the WAWI website declared that the northern-hemisphere WAWI partners would 'engage with national and local governments, citizen groups, and communities in West Africa' (Water for the Poor 2005a).

This implied that African stakeholders, both governmental and non-governmental, were to be integrated into WAWI's activities. CNHF documentation explicitly stated that as well as northern hemisphere participants, equally 'important *partners* are the national governments [of Mali, Niger and Ghana] and local communities which will provide some financing and valuable in-kind assistance' [emphasis added] (CNHF 2004a).

Implementation and Partnership Modalities in Practice

One of WAWI's key objectives is to 'foster a new model of partnership and institutional synergy to ensure technical excellence, programmatic innovation, and long-term financial, social and environmental sustainability in water management that may be replicable elsewhere' (CNHF 2004a). A WAWI document (Doyle & Corliss 2006, 1) stated that 'Field staff and Headquarters' representatives of WAWI partners are all agreed that there have been important benefits to collaboration' in WAWI. One interviewee (15, 2004) judged that WAWI was growing into a genuine partnership: 'based on the introduction that's been made between some of the partner organizations, they've gone on to work together on other projects . . . the partners are coming together and collaborating on other activities that go beyond WAWI which suggests that a real sense of partnership is being nurtured.' Many of the organizations involved had worked with each other on previous collaborations. For example, before WAWI came about, World Vision was already working with UNICEF and WaterAid (Interviewee 28, 2005), while Lions Club International had worked with the Carter Center, HKW, USAID and WaterAid on other projects (Interviewee 29, 2005). One WAWI partner representative (Interviewee 27) reported in 2004 that through WAWI, HKW and WaterAid were in discussion about forming a partnership 'to jointly seek funding for water projects in areas that WAWI doesn't cover . . . the idea being that . . . [WaterAid] have water capacity and expertise that . . . [HKW] don't, [and HKW] . . . have trachoma control capacities and programmes that need water.' Another WAWI partner (Interviewee 23, 2005) reported that being involved in WAWI has 'given us introductions to relationships with people we wouldn't have otherwise spoken with, or it's given us relationships that are more intimate than we would otherwise have managed.'

Although some partners within WAWI, such as World Vision and the CNHF, contributed greater amounts of capital than others, a representative from a partner organization committing fewer resources than these two organizations (Interviewee 24, 2004) stated that 'they treat us as equal partners. Everybody has different expertise to bring to the table, so even if you're not fronting the major costs of the initiative you're still considered as being high up there.' Similarly, another representative (Interviewee 27, 2004) from one of the partners contributing little or no direct funding to the initiative, maintained that 'we certainly have an equal say (or have a good say at least) in the operations and direction of WAWI.' However, another partner representative (Interviewee 25, 2004) claimed that, in relation to the CNHF, 'Everybody else falls into line underneath them. Even though it's a very collegial relationship, they are the 900 pound gorilla in the partnership.' Another participant (Interviewee 26, 2004) agreed, stating that:

> Hilton and World Vision dominate because they are, I'd say, 80 per cent equity holders in the project, so they bear more listening to than the others . . . I think [all partners] share a perspective of partnership, but . . . I suspect for most organizations they probably don't feel their boat bears equal weight with all the others.

At the level of implementation on the ground, World Vision has a central role, if not a dominant one:

> World Vision are the flag wheel in terms of the actual on the ground work . . . it's because of their [existing] programmes that the three countries were chosen . . . they are the ones that we all look to at one time or another for some sort of logistical support and assistance . . . By default they are the co-ordinating body . . . they're the main implementing partner, there's no doubt about it, it's really through them that things get done (Interviewee 25, 2004).

Another headquarters partner representative (Interviewee 29, 2005) said that 'A lot of the activities hinge on them [World Vision].' One interviewee (27, 2004) asserted that 'World Vision clearly has the leadership in terms of contributing the most financially and I would say that there is also a proportionate amount of control that they have within WAWI . . .

they are . . . the principal player.' Indeed, another interviewee (28, 2005) reported that of the total resources of WAWI, 'about 80–85 per cent . . . are being used by World Vision at the field level.' This interviewee (28, 2005) suggested that World Vision tried to 'play down' the fact that they were 'the lead agency in each of the [WAWI focal] countries . . . because to have a good partnership you have to make sure that no one particular partner dominates.'

A further source of tension within the partnership is that the project locations chosen by WAWI are largely areas in which World Vision were already working, but some of the other partners, such as HKW, wanted to see WAWI's areas of operation expanded to include their own existing focal locations in the three WAWI countries (Interviewee 27, 2004; Interviewee 29, 2005). One participant (Interviewee 26, 2004) reported that this had 'been a bit of an area of contention because some organizations have lobbied to move WAWI beyond its existing areas because they rightfully say that the needs are greater elsewhere.' However, 'there is a reluctance on the part of the principal members to . . . [expand the WAWI territory] because there is so much to do in the areas they're already working in' (Interviewee 27, 2004).

Moreover, World Vision's geographical focus made it difficult for some WAWI partners to operate at all. For example, one WAWI member (Interviewee 23, 2005) said that although the 'ITI would like to be an implementing partner . . . it's struggling to find a project area that overlaps with one of World Vision's in terms of geography:'

> Conrad Hilton definitely had this vision of bringing together lots of different agencies with different skill sets, to work in a specific area to deliver services in that area. I think that may be partly their philosophy, but it's also modelled on the relationship they've had with World Vision, which is very much geographical – that's the way they work, on a geographical basis, For organizations that are sectoral, like the International Trachoma Initiative, UNICEF and WaterAid, that has caused some difficulties.

However, according to another WAWI participant (Interviewee 25, 2004), WAWI was improving in its partnership synergy over time:

> My sense is that we are becoming more effective as time goes on.

> We didn't really know who we were as a partnership early on, so any effectiveness then was probably just a by-product of each individual organization's initiative . . . work wasn't really getting done within the context of the partnership . . . that's changing now, there has been a lot more discussion . . . about sharing resources [within a country] . . . That's the sort of stuff that is really important to the Hilton Foundation – to be able to see that kind of cross-collaboration . . . One of the [other] things we're trying to do is coordinate fundraising activities because every organization has their own . . . fund generating protocols . . . [and] strengths . . . so we're trying to leverage and coordinate those so that we can target specific needs for WAWI, and address those needs through the partnership's collective resource generating capabilities.

Similarly, another WAWI partner (Interviewee 28, 2005) reported that 'synergies are well established at the field level' because each partner concentrates on addressing the issue they are working on without interfering in other partners' work, 'but at the same time co-ordinating' with each other, especially where 'interventions overlap'.

Nevertheless, some partners, when interviewed, remained unconvinced that WAWI had developed any significant synergistic qualities. For example, one partner representative (Interviewee 23, 2005), while conceding that 'at a local level there has been a reasonably good amount of exchange of experience and contact between the different partners,' asserted that 'There was certainly a sense when the initiative started off that this should be something about how we can combine the talents, the abilities, and capacities of these different organizations together, into some sort of greater whole. I'm not absolutely clear that we've done that effectively.' Another interviewee (29, 2005) maintained that this was 'one of the big challenges' the partnership faced, but that rarely did all the members within one country work together in an integrated fashion: 'if we're talking about field-level activities then I don't think that more than two or three [organizations work together on something] at a time . . . I think we're struggling with the fact that people are working in different areas and people have different organizational cultures about what they can and can't do.'

The fact is that there were fundamental divergences in outlook between the different partners, especially in their perspectives on development, and in how they perceived their roles in promoting it. As one WAWI partner representative (Interviewee 23, 2005) explained:

> different parties to it, bring different ideas of what it's for . . . the Conrad Hilton Foundation is very much coming from a philanthropic perspective and is about providing scarce resources to help more people get access to safe water . . . World Vision would very much be in the Conrad Hilton camp – they have development areas and development programmes where they work long-term for 15–20 years, looking at a whole cross-section of development issues, of which water and sanitation is only one, and I think they have a core interest in accessing resources to develop their project area . . . the USAID comes to it from a more strategic perspective in asking 'what is it that this group of people can do together in the three countries to have a broader impact on the water sector?' Which would also be the perspective that WaterAid would normally come from too . . . it would be aiming to do projects which have impacts on beneficiaries, but to use those projects to lever change in the sector in the countries it works in . . . WaterAid and UNICEF are much more interested in a sectoral approach to a country, as opposed to working in specific areas . . . the International Trachoma Initiative is probably similar to UNICEF and WaterAid in its approach. Then you have the Desert Research Institute and the Carter Center who have a slightly more academic interest in the whole process, so it's very different again.

Those with a strategic outlook wanted WAWI to assume a policy influencing role that would have an impact on the water sector as a whole within each country, and possibly beyond, because, as one WAWI partner (Interviewee 27, 2004) asserted, 'when you bring a critical mass like WAWI . . . you have a better chance of getting the attention of governments at all levels, local, regional, and national' than if one organization tried to influence them independently. Another partner representative (Interviewee 23, 2005) argued that:

> somehow we should be trying to make use of the entirety of the work of the key partners [not just their work within the scope of WAWI] to lever change in the sector . . . using our work as a

> demonstration of ways of solving . . . blockages in the sector and then using that to lever change with other major donors and other nationally funded programmes . . . and maybe change national government attitudes, policies and practices . . . We have an opportunity, with quite a reasonable sized chunk of work in each one of the three countries, to actually use that work for a broader strategic purpose, and I feel it is incumbent on us to try and do that . . . At the moment it's more about delivery on the ground and working with the communities . . . It's focused on outputs and beneficiaries. That's not bad in itself for an initiative; at least it's getting services to people which a lot of the larger initiatives don't end up doing. I think my only argument is that it could do more. It could do that and try to do that in the context of the main problems in the water sector in Ghana or Niger or Mali, and therefore try and make a bigger difference.

This interviewee (23, 2005) claimed that World Vision did not supply this strategic leadership:

> the organization that has been charged with hosting the Secretariat is principally interested in the basic outputs, and not in the strategic impact (or feels uncomfortable with that), because that's not the role that they have traditionally played . . . If we were to treat this as a serious initiative . . . we'd have to look again at the Secretariat . . . because you would need to have a . . . Secretariat that was interested in the potential broader strategic impact that the initiative could have . . . .if it was going to be more strategic I think it would have to develop a different way of working, it would have to be a much more active set of relationships because we would have to be trying to work out what all the common lessons and ideas are that are coming out of the work that we're doing that we can all agree on. And we'd also have to have some sort of plan . . . a campaign of action . . . to try and change the way other people do things, and that would mean that we would need to know what it was that we wanted to say, who we were going to say it to, and who was going to take responsibility for saying what to whom. That would be quite a feat, to actually present a united face, across very disparate organizations, to an external body.

The fact that different partners had such different approaches and distinct ways of working raised the question of whether WAWI partners could work as a collective whole, or whether they could only concentrate on their own specific projects. As one partner (Interviewee 23) put it in March 2005, was there an overall 'strategic view' to the initiative, or was WAWI just 'a ragbag collection of donors, funding a ragbag collection of projects, with no coherent central theme.' His conclusion was that although there had been some discussion about 'developing a strategic approach,' the initiative remained, 'in practice, a collection of individual projects . . . funded by a different combination of donors . . . under the common banner of WAWI.'

These different partner stances also manifested themselves in the differing views partners took on the role of indigenous government in the initiative. In its early days, WAWI tended towards northern domination because it began life as a northern partnership, with the CNHF building relationships with other organizations whose head offices were located in the northern hemisphere. Since then, one of WAWI's stated ambitions has been to work in partnership with indigenous governments, but this ambition has not always been fulfilled. One partner (Interviewee 23, 2005) insisted that:

> It is a northern partnership which works in the south . . . I don't think [African governments] . . . are involved and that's why I think it's a principally northern partnership . . . I think there have been formal meetings with government at various stages but [indigenous] government is not an active partner in any true sense of the word.

At the end of 2004 (two years after work began in Mali, and a year after work began in Ghana and Niger), a WAWI participant (Interviewee 24, 2004) maintained that 'the hope is that eventually we partner with the governments in the countries . . . I know in Ghana there has been some interaction with the government . . . but . . . it is the beginning phases of getting them on board.' However, this partner argued that despite the fact that the co-operation of African governments 'would help in the long-run to make sure people continue to do the things they have learnt,' and that partners 'can go in and help for so long, but eventually someone is going to need to take over,' a lack of indigenous government backing would not 'impact on the partnership directly because . . . they . . . [were not] contributing financially.' This suggested that the collective belief in the need to partner with indigenous government was not taken seriously by at least some of WAWI's implementing organizations.

Another WAWI participant (Interviewee 25, 2004) held that while it was important to work with indigenous governments, and WAWI made every effort to do this, they were still not considered partners:

> in each of the countries there has been a very conscious effort to engage governments. Whilst we've not tried to make them formal partners, we have tried to have them embrace WAWI and participate . . . whether it's data sharing, whether it's with providing some sort of logistical help . . . and those sorts of things. There is an ongoing effort to keep the governments involved that is facilitated, to a large degree, by World Vision's activities . . . who I think recognize the fact that life's a lot easier if you're working with the various ministries and so on.

However, one WAWI partner (Interviewee 27, 2004) disagreed, maintaining that African governments were, in fact, partners, claiming that, 'we do have a partnership with the African countries, we would not be able to do anything that we're doing if we didn't have the blessing and active collaboration of the governments in the three countries where WAWI works, that's the foundation of what we're doing.' Another partner representative (Interviewee 29, 2005) took the middle ground between these two views, stating that 'I wouldn't say that [African governments and their agencies are] . . . not partners, but at the same time they're certainly not sitting around the table at headquarters level. They are more involved at the country and local level.'

This is not to deny that WAWI was heavily involved with indigent governments. On the contrary, World Vision's ADPs in Ghana, which were the main areas in which WAWI operated, 'correspond to the [local governmental] Districts,' and World Vision worked closely with the District Assemblies because 'those local relationships are key to getting work done' (Interviewee 15, 2004). Furthermore, one participant (Interviewee 26, 2004) argued that at both national and local levels efforts were made to work with government: 'WAWI works very closely with government organizations to determine their priorities,' and the areas which WAWI is working in 'were decided upon in collaboration with the government' of each country, because:

> you can't really do anything in these countries without the approval of the government . . . In Ghana . . . there is a strong local level element . . . [but in] the other two countries it's more at the national level because the local level authorities aren't really up to that yet . . . [but] at the national level it's more of a struggle to keep them engaged (Interviewee 26, 2004).

However, indigenous government was more engaged in sanctioning WAWI's work, than in implementation at the field-level.

This issue again comes down to the different perspectives assumed by different partners. Each organization's outlook on the aims of the initiative, and on their own work, influenced the view they held about the role of indigenous government. World Vision's perspective was hands-on: 'World Vision tends to have lots of their own agents, and so focus . . . on service and delivery . . . World Vision . . . own the rig, their people are going out drilling boreholes, their trained agents are the ones delivering health and sanitation education in the villages' (Interviewee 29, 2005). On the other hand, the perspective held by organizations like the Lions Club, HKW, the Carter Center, and UNICEF was capacity building, 'working through existing government structures:' These organizations 'may have technical people working with the Ministry of Health [for example, so] . . . they're still delivering services . . . but it is through somebody else, through training somebody else and helping them do it' (Interviewee 29, 2005). A WAWI partner (Interviewee 15, 2004) explained how, for instance, UNICEF in Ghana were 'working both with District Assemblies in the north, and with national ministries in the capital, both specifically with regard to their role supporting the West Africa Water Initiative, and more broadly to build their capacity and policy regarding rural water and sanitation.' A headquarters partner representative (Interviewee 29, 2005) explained that 'this is a struggle that a lot of us in the partnership question – what is the best way to go about it' – to build the capacity within existing structures, or concentrate on actually delivering services.

This distinction should not, however, be exaggerated, because even in the work that centred on delivering services directly to communities (which was consistently the main focus of WAWI because World Vision were the lead agency on the ground), there was an element of skills and knowledge transfer. There was also a long-term presence within a community: 'they're not just drilling a borehole and leaving' (Interviewee 15, 2004). As one interviewee (25, 2004) explained:

> there's a large training component to what we do so . . . when we leave, they can still continue to implement the technologies we've trained them in. Training people in Africa is a huge part of what we do . . . we have to demonstrate that we're doing that . . . It's on the ground work [that is the focus] but I think all of the partners are interested in, and dedicated to, building local capacity.

In terms of civil society involvement, a WAWI partner (Interviewee 27, 2004) asserted that 'their collaboration, their participation, their buy-in is absolutely essential. But we're working in some areas where you really don't have much in the way of organized civil society stakeholders . . . some of the places like in parts of Niger – there's not much out there.' Another interviewee (15, 2004) reported that 'there are ways in which people are given options in terms of how they want to participate in the work that is being advanced by WAWI,' but 'there isn't a universal approach . . . each organization has a somewhat different way of going into communities.' World Vision certainly:

> engages community leaders . . . World Vision do not decide where a borehole is going to be. Ultimately the community and community leaders help with that . . . and that relates a lot to the desire to have the water project be sustainable and doing what you can to make people real stakeholders in each well or borehole, which includes, for example, the building of community-based teams who know how to repair wells and know how to get parts for it and things like that (Interviewee 25, 2004).

Furthermore, many of the field staff of northern-based partner organizations were local residents: indeed, one interviewee (26, 2004) suggested that as many as '98 per cent of the . . . field agents . . . who are drilling the wells, providing hygiene education, and all those sorts of things . . . are local people.' Another partner (Interviewee 15, 2004) maintained that 'World Vision Ghana is staffed, by and large, by Ghanaians.' The WAWI website (WAWI 2006) reported that often WAWI partners use 'indigenous private and non-profit organizations in the country' to 'implement . . . [their] activities on the ground.' This was certainly true of WaterAid, who put the 'bulk of funds' received from the CNHF 'through . . . local partner organizations,' the 'primary local partner' participating in WAWI being New Energy (Interviewee 15, 2004).

Moreover, one northern partner (Interviewee 15, 2004) claimed that, although WAWI originated in the north, and many of the northern partners 'have the perspective that we know that there are some general categories of goals that we want to move toward . . . in terms of specifically defining those goals and how to get them, we're looking to the people in West Africa to determine that . . . leadership needs to come from the field.' Similarly, another headquarters partner representative (Interviewee 29, 2005) stated that 'It started more as northern but I think that . . . all the northern organizations involved really wanted to put the focus on the field-level and the grassroots-level, which has made it more of a north/south partnership in terms of ideas coming from the field . . . that say "this is what we want our objectives to be".'

Evaluation of Outcomes

Moving to an appraisal of the outcomes of WAWI's work, we found an immediate difficulty – that of obtaining information. Although the WAWI initiative professed to have 'proceeded through all the normal stages of partnership evolution,' and therefore had 'significant lessons to share about what it takes to make a complex partnership work' (WAWI 2006), these 'lessons' were not made available in the public sphere. The WAWI website (headed *Water for the Poor*) is our main source of information about the partnership. Together with the CSD database with its voluntary reporting mechanism, it is one of the primary vehicles for ensuring that the partnership operates in a transparent fashion, providing it with external accountability and legitimacy. However, after its creation in 2002, the website was not comprehensively updated until the end of 2004/beginning of 2005. It was updated again in early 2006, but that update only provided a new list of visits and meetings that had taken place over the last year, with no minutes.

Moreover, we found that obtaining partnership documents from WAWI partners was very difficult, because this information was rarely released into the public domain. A WAWI partner acknowledged that 'I think we have done a good job at getting the work done but we haven't done a really good job at communicating it outside of our little network' (Interviewee 24, 2004). More importantly, there was no comprehensive, overall monitoring and evaluation strategy in place. As one partner (Interviewee 23) noted in 2005:

> The interesting thing is that you have this initiative which is, in a sense, a body in its own right, yet each individual partner still has individual funding relationships with the various donors. We all report separately . . . we aren't part of any corporate report . . . so it's left to the individual partners to provide their means of verification for their particular outputs, to their particular donors.

Nevertheless, there are some reports on WAWI's outcomes. For instance, in the 2004/5 website update, Nerquaye-Tetteh (2005) stated that:

> the WAWI partners are rapidly fulfilling . . . [their] objectives as shown most prominently by the following:
>
> - Wells being drilled
> - Schools hygiene being promoted
> - Government institutions being strengthened
> - Community sensitization occurring in water and environmental sanitation
> - Vegetable gardening being developed through small irrigation systems in communities.
>
> In addition, and most importantly . . . [WAWI will soon] begin the process of registering positive impacts in management of guinea worm, trachoma and diarrheal diseases, especially among children, and overall improved community health.

Also on the WAWI website there was a detailed breakdown of objective achievements for each country, showing that: boreholes had been dug; latrines, waste pits and refuse disposal sites had been constructed; communities had given in-kind contributions; research studies had been done; training in improved health and sanitation for teachers, students and communities had been undertaken; education materials had been created; school health clubs had been established; communities had been encouraged to form enterprise development groups to manage water systems and management committees; men and women had been trained in the maintenance of water points; village banks had been created to manage micro-credit activities; women's associations for micro-credit projects had been set up; water testing had been done; micro-irrigation demonstration and pilot sites had been established; farmers and trainers had been trained; WAWI partners had been trained in hydro-geological management; and gender mainstreaming workshops had been held for WAWI partners and government officials to harmonize gender approaches (Water for the Poor 2005b; Water for the Poor 2005c; Water for the Poor 2005d).

According to the CSD partnership database (WAWI 2006), by February 2006, WAWI had accomplished the following:

- 330 boreholes drilled and 110 village water points rehabilitated in rural areas, with community management committees trained, and mechanisms for financial sustainability in place
- 11 peri-urban water points installed (Ghana and Mali)
- 8200 latrines constructed with community training in hygiene education
- School-based sanitation programs underway in core WAWI geographic areas
- 13 pilot areas developed for micro irrigation and gardening, including training of trainers, and developing market linkages (Ghana and Mali)
- Hygiene promotion, gender mainstreaming, GIS [Geographical Information System], and IWRM workshops held for all WAWI partners, government, and collaborating institutions
- WAWI engagement in national sector planning, including Mole (Ghana) and SIDEAU (Mali) conferences, participation in working groups on school sanitation, blindness prevention, and water/sanitation sector coordination.

However, even though these outcomes were achieved, as one participant (Interviewee 26, 2004) pointed out, 'a well by itself, or a latrine by itself brings a certain measure of change, but hygiene is the factor that really changes things, so it's all about behaviour changes which is more challenging.' So it will be years before the sustainability of WAWI's impact becomes clear.

Finally, the question arises of the significance of such outcomes to the WAWI project. As discussed above, discord existed among partners regarding the overall purpose of the initiative – whether it was essentially about immediate outputs such as the delivery of services to beneficiaries; or whether it should have a strategic focus, aiming to utilize the work of WAWI partners in a broader way to influence the water and sanitation sector as a whole, within each country and perhaps beyond. This tension worried many partners, with one interviewee (25, 2004) complaining about the lack of a strategic plan for the initiative: 'We're a couple of years down the road and we still don't have one.' However, in 2005 a strategic planning process was initiated for the partnership, and was completed in January 2006 (Doyle & Corliss 2006). Funded by USAID, managed through ARD, and undertaken by the Strategic Planning Working Group and the WAWI Secretariat, with the assistance of some consultants, the plan was created: 'as the thirteen partners worked together, they were motivated . . . to sharpen the Initiative's focus and clarify near-term, medium term and long term commitments . . . with the view of scaling-up a successful model of partnership within current countries, the region and potentially other regions' (Nerquaye-Tetteh 2006), and 'ultimately significantly expanding . . . [the partnership's] reach and influencing sector policy and other investment decisions.' Lack of a strategic plan in the past had meant that WAWI suffered 'from some significant challenges that hinder[ed] its effectiveness – including fragmented plans, missed opportunities for cooperation and minimal documentation of collaborative successes' (Doyle & Corliss 2006, 1). The plan listed three strategic priorities (Doyle & Corliss 2006, 2):

1) Creation of a coherent integrated operating model at the local, country and regional levels (including geography and service mix)
2) Creation of partnership synergy beyond the initial core project areas . . .
3) Effective engagement of WAWI within an evolving international context

The general consensus from partners was that if the strategic plan were implemented successfully, in five years time, WAWI would be able to:

- Produce results that are greater than the sum of the individual Partner efforts through strengthening partner commitment, coordination and an effective and representative Secretariat;
- Take advantage of the growing global awareness of the importance of and willingness to invest in water and sanitation;
- Develop a proven and documented model for collaboration that can be adapted to address critical water and sanitation needs elsewhere in the world;
- Strategically use Partners' critical 'mass' of experience to communicate with and influence the water and sanitation policies of governments, development banks, donors, IOs [International Organizations] and NGOs at local, national and international levels; and
- Exhibit strong fundraising capability and secure funding for all WAWI Partner activities to support the scaling up of the integrated operating model within an expanded service area (Doyle & Corliss 2006, 2).

A partner representative (Interviewee 27) optimistically declared at the end of 2004 that:

> If we continue to work collaboratively then I think we can do more together ultimately, than we could if we were working independently. Actually, I think that if the WAWI model survives and we can figure out how to really maximize the benefits and get past all the little humps and bumps, then WAWI could be a model for development in other geographic areas in the future . . . at the very least WAWI will produce another set of lessons learnt – here's what we tried to do, we had money, we had the right technical expertise, and if it didn't work then we need to look at it and ask why not? . . . but I certainly think that we've got a good chance of doing better than that.

Conclusion

In this chapter, we have analysed the third case study – WAWI – paying particular attention to its origin, structure and working. In our evaluation of its work, we found that its partnership credentials were ambiguous, in that although it espouses an inclusive model, in practice, the level of its engagement with indigenous governments and civil society has been limited. On its substantive role, WAWI originally concentrated largely on the direct delivery of water and sanitation services, but later it became more interested in the possibility of establishing a strategy to use its direct delivery work to contribute to the reform of water sector governance. In the next chapter, we compare and contrast WAWI with the other two TTPs – EUWI and PAWS.

6

COMPARATIVE ANALYSIS OF THE THREE CASES

Introduction

In this chapter, we reflect on the three case studies and make comparisons between their processes of partnership, their working styles, and their achievements, drawing on the conceptual framework of partnership developed in chapter two. We begin by examining the cases in the light of a typology which distinguishes three different, though not exclusive, overarching strategies for TTPs for drinking water and sanitation in sub-Saharan Africa: 1) reform of sector-wide governance; 2) capacity building within existing governance structures; and 3) direct delivery of services. Next, we consider how far the three TTPs have met the 13 criteria for partnership processes set out in chapter two. To do this, we first compare the ways in which the three partnerships have accommodated the claims of different stakeholder groups, and we address the critical issue raised in chapter two about the TTP framework: whether it is an attempt to off-load governmental responsibilities to other actors, resulting in a decline in the public accountability of decision-makers; or whether it follows the model of the UN Commission on Sustainable Development (CSD), whereby non-governmental actors are engaged with, but governments retain their control over decision-making processes.

Second, we assess how far the three TTPs have overcome inequitable relationships between the north and the south, and established in-country ownership. Here we focus on southern governmental participation because, although the involvement of other stakeholders in the south is important, governmental buy-in and active participation is vital to ensure the sustainability of any developmental strategy aimed at achieving the MDGs. This section also discusses the impact that the developmental practices originating from the northern-hemisphere base of each partnership have had on its approach on the ground. We then move on to consider how far the three TTPs have met the nine criteria for partnership outcomes set out in chapter two, and finally, we conclude the chapter with a summary suggesting that the full potential of TTPs has not yet been fulfilled.

Partnership Approaches

The three partnerships examined in this book are linked by their common goal of working to improve water and sanitation services in sub-Saharan Africa, thereby contributing towards efforts to meet the MDGs in this area. However, as we have learnt, these partnerships differ in their structure, their partnership arrangements and processes, their specific objectives, and their means for achieving their objectives. One of the most useful means to differentiate between the partnerships is to observe their overall strategy within Africa, which we can classify into three different approaches: *reform of sector-wide governance*; *capacity building within existing governance structures*; and *direct delivery of water and sanitation services.*

By *reform of sector-wide governance*, we mean devising a strategic plan to improve the governance of a water and sanitation sector as a whole in an African country, identifying gaps and problems, and making recommendations on how to address them. By *capacity building within existing governance structures*, we mean working within existing structures, determined by government, to foster competence, and provide the necessary skills to individuals and organizations to enable them to function more effectively and efficiently in their given role, and to ensure that improvements are sustained once donor organizations leave. In other words, capacity building is 'An approach to development that aims to instil commitment and improve fundamental management and technical skills within an organization [including community organizations or groups], thereby making the institution more effective and sustainable' (COMPASS 2006). According to Biermann *et al.* (2007a, 256), most partnerships are mainly concerned with 'institution-building' rather than 'implementation'. By *direct delivery of water and sanitation services*, we mean projects that are on the ground, providing safe water supplies and latrines. Of course, the sector-wide governance and capacity building approaches are ultimately aimed at the improvement of water and sanitation services, but not directly. Here the approach directly provides such services, or builds operational systems for the direct delivery of these services.

Each of the three partnerships has inclined towards one of these approaches: EUWI inclined towards sector-wide governance; PAWS inclined towards capacity building; and WAWI inclined towards the delivery of services. However these respective strategies are not mutually exclusive, and the boundaries between them are hazy: none of the partnerships studied fixed upon only one of the three approaches to the entire exclusion of the other two approaches. Instead, it is more accurate to say that the centre of gravity for each was different, with more or less emphasis given to the favoured approach, but with the emphasis shifting at different junctures of the partnership's history, which we now trace.

EUWI (EU 2002, 4), stated that its focus was neither on capacity building nor on direct service delivery (since it had no funds of its own), but on the improvement of governance, since it hoped to:

- Reinforce political will and commitment to action . . .
- Make water governance effective and build institutional capacity . . .
- Improve co-ordination and co-operation . . .
- Increase the efficiency of existing EU aid flows . . .

Through reinforcing political will, EUWI aimed to work at the highest level to catalyse sector change, by means of harmonized donor assistance and improved sector co-ordination, co-operation, and efficiency. These are all ingredients of a strategic effort to enhance governance arrangements which EUWI sought to implement through a country dialogue approach in Africa. The country dialogues were two-phased: a first phase to brief relevant parties on the initiative and develop an action plan for the country through dialogue; and a second "action" phase. This approach was taken in Zambia (the focal case examined in this study for EUWI). However, it proved to be too vague, lacking results specifications, and in Zambia the second, or "action", phase never began. Therefore, by 2005 a new direction for the dialogues was developed, whereby the dialogue process would result in a roadmap for each country, indicating a long-term strategy on how it intended to reach the MDGs and how this would be funded. In this strategy, annual targets and outputs would be in place, linked to the inputs required to meet them. Through engaging donors, governments, and non-governmental stakeholders in the planning process and its implementation, the logic was that governance as a whole would be enhanced, as in-country ownership of the plan was improved and wider perspectives were taken into account, and this would ultimately result in more sustainable solutions. This would entail capacity building work, and/or service delivery projects, though we have not seen any country dialogue reach this stage.

The stated aims of PAWS (2006a) were wide-ranging:

- To deliver tangible and sustainable benefits . . . at the local community level.
- To emphasize strong inter-linkage between water supply and sanitation.
- To encourage best practice and develop guidelines for effective and sustainable tri-sector partnering.
- To focus attention on secondary towns and peri-urban areas . . .
- To demonstrate the effectiveness of multi-stakeholder engagement.
- To promote socially and environmentally sustainable services that endorse principles of good water management
- To champion an environment that attracts investment and ensures financial sustainability.

Like EUWI, PAWS had no funding of its own (beyond that given by the UK government for running costs). Thus, although at first PAWS gave the impression that it was aimed at service delivery, it soon became clear that its driving focus would be capacity building and knowledge transfer within existing government structures. This took many forms, including support for regulation, socio-economic strategies, policy and institutional assistance. In South Africa, the majority of PAWS's engagement centred on capacity building at the level of local government, partly because this was an area of need identified by the South African government which PAWS could assist with, and partly because PAWS wanted to work near the ground. The partnership did, however, do some work at the national level, offering advice to the national government on particular water supply and sanitation issues, and more recently, providing support on water resources management at a strategic level. Because the South African government was in the process of implementing new strategic frameworks which were not yet operational at the local level, this was the platform in need of support, and since PAWS was able to offer such assistance once the Environment Agency joined the partnership in early 2004, it did so. In Uganda, PAWS worked with the central government to build capacity around regulation issues, and agreed a programme of wide-ranging capacity building which entailed working with civil society groups and the private sector. In Nigeria, however, PAWS took a different tack, moving closer to a service provision type of partnership, with project work undertaken at the community level to set up new water delivery structures. However, capacity building in existing government structures was still at the heart of its approach, since the aim was essentially to provide technical support to the Nigerian State of Benue, and help them to establish community-run pilot schemes.

WAWI was different from both EUWI and PAWS in focusing on directly increasing water and sanitation services. In its three focal countries, it aimed (CNHF 2004a):

1) To increase the level of access to sustainable, safe water and environmental sanitation services in rural and peri-urban communities
2) To decrease the prevalence of waterborne diseases including blinding trachoma, guinea worm and diarrhoea
3) To ensure ecologically and financially sustainable management of water quality and quantity; and,
4) To foster a new model of partnership . . .

Indeed, WAWI had specific numerical targets for delivery, including 865 boreholes, 150 alternative water sources, and at least 9,000 latrines (CNHF 2004a, WAWI 2006), and each implementing partner had its own contractual agreement with its funders, to achieve its own targets within this overall framework. However, if service delivery was the primary focus of the partnership, capacity building was its secondary focus. As interviewee 26 (2004) said, 'Capacity building is a very important component of WAWI.' This capacity building occurred on three levels. First, field staff from some partner organizations were trained in skills by other partners (Interviewee 26, 2004). Such capacity building work was clearly outside government structures. Second, capacity building work was carried out by World Vision (the lead WAWI implementing partner), which had a strong focus on hardware installation but also trained communities in pump maintenance and repair, and latrine and laundry pad construction (CNHF 2003). Third, UNICEF's role in the partnership was to 'support capacity building of rural water supply programmes at the national and sub-national levels.' WAWI also held workshops on hygiene promotion, gender mainstreaming, GIS, and IWRM, which some government officials attended. There were calls from some partners to give the partnership more of a strategic focus, whereby the project work that WAWI partners were carrying out could be used to influence the sector as a whole, and some work was undertaken to achieve this goal, but although this showed an inclination towards the strategic sector-wide governance approach, it was only a tentative move in that direction.

Although EUWI showed the greatest inclination towards the strategic sector-wide governance approach, all three partnerships demonstrated a tendency towards governance improvement in the water and sanitation sector in the countries within which they worked, albeit in different ways, ranging from what one might loosely term a bottom-up to a top-down approach. EUWI took a top-down approach in that it sought to work at the highest political level to create a strategic plan for the improvement of governance of the water and sanitation sector in a country, although it did (at least theoretically) engage stakeholders in this process. WAWI came from the opposite direction, employing a bottom-up approach towards governance improvement, not in the sense that it empowered African citizens to make demands on their governments, but in the sense that its core activity was at the ground level, providing services and increasing the capacity of individuals and communities to provide and maintain their own services, thereby influencing the sector, and improving governance from below. PAWS was situated between EUWI and WAWI in its approach to governance: neither top-down (like EUWI working with national governments to determine strategies for improving sector governance as a whole) nor bottom-up (like WAWI using its project efforts to influence sector reform), but instead working with governments to deal with specific capacity issues (institutional, technical, or human) on which support was needed, and thereby contributing to improving sector governance.

Interestingly, one interviewee (23, 2005) saw EUWI and WAWI travelling from opposite directions to meet in the middle: although EUWI started as a strategic governance initiative, 'actually in some ways it has ended up like a massive WAWI in the sense that . . . from attempting to be . . . strategic . . . with donor coordination and harmonization, it's actually turned into a project fest. So maybe where WAWI and it meet is that they've come to the same point from different directions'; while WAWI began life doing projects but later assumed a strategic outlook. Although this interviewee's interpretation is inaccurate in that it included the EU Water Facility (which funded projects) as part of EUWI, it did illustrate the fact that despite their essentially divergent perspectives, there were important overlaps between the different TTPs.

In a sense, none of the TTPs could avoid having a strategic dimension, because the very nature of a TTP entailed reform of governance, both within Africa, and within the northern context, in that it required that stakeholders beyond government were involved in policy making processes and implementation. This requirement was set out in principle two of the four principles of water management (which became known as the Dublin Principles) established at the UN's International Conference on Water and Environment in 1992 (Global Water Partnership 2008):

> Water development and management should be based on a participatory approach, involving users, planners and policy-makers at all levels. The participatory approach involves raising awareness of the importance of water among policy-makers and the general public. It means that decisions are taken at the lowest appropriate level, with full public consultation and involvement of users in the planning and implementation of water projects.

As one interviewee (3, 2003) maintained, through their multi-stakeholder nature, TTPs have the potential to 'mainstream principles like transparency and consultation . . . into governance, into general government working. So in that way they can . . . hopefully ultimately ensure better decisions are made because more people's views are taken into account.' If the TTP framework mainstreams such principles of participation into existing governance structures, regardless of the concrete outputs of the partnerships, their very existence is important as a governance improvement tool. In the next section, we investigate how accommodatory in practice these TTPs have been to stakeholder participation.

Multi-Stakeholder 'Partnerships'?

In this section and the next section we consider how far the three TTPs have met the 13 process criteria for partnerships set out in chapter two. These criteria are as follows:

1) Guarantee legitimacy, through agreeing the design of the partnership 'in a democratic, transparent and equitable manner, including the identification of stakeholder groups and participants, the framing of agenda and work plan' (Hemmati 2002, 60).
2) Involve 'all partners [or at least a good range of partners] from the outset (rather than the traditional sub-contracting approach)' (UNCSD Secretariat 2005).
3) Be 'inclusive and not exclusive' (Hemmati 2002, 59). If logistical and functional constraints make selection criteria necessary then these criteria need to be made public and open to debate.
4) Ensure there is a broad forum of stakeholders, in some cases involving 'those not necessarily part of project implementation but interested in or impacted by the partnership project' (Caplan *et al.* 2001).
5) Clearly define roles for all stakeholders which allow for flexibility, but have clarity about 'who is responsible for what and when' (Evans *et al.* 2004, 14).
6) Realize a 'common understanding of the intent and outcome of the process' (IDSUH, IDSRU & CMI 2002, 8).
7) Be willing and have the capacity to learn, evolve and adapt to changing circumstances.
8) Hold regular meetings and build transparent and 'solid decision-making structures' which embrace horizontal as opposed to hierarchical coordination and accountability (Caplan *et al.* 2001).
9) Ensure equitable participation of all partners, or at least an environment conducive to equitable participation even if one or more 'partners' choose not to engage as fully as others do.
10) Involve intended beneficiaries, avoiding top-down prescriptions.

11) Attain mutual understanding of individual partner organizations motivations and constraints (Evans *et al.* 2004, 14).
12) Achieve an appropriate balance made between mutuality and autonomy.
13) Create and maintain good communication channels, both internally to build relationships between partners, and externally to demonstrate effectiveness, attract new partners and funding, and improve legitimacy (UNCSD Secretariat 2005).

Ever since the Rio Earth Summit in 1992, international sustainable development meetings have increasingly encouraged the participation of non-governmental stakeholders, although usually to give them 'voices not votes' (Hemmati citing Edwards 2002, 61). At WSSD, the advent of TTPs marked a step forward in this participatory process, giving non-governmental stakeholders a more prominent role in the implementation of international governmentally agreed goals. As Mitchell (2003) claimed, 'one fundamental thing from Johannesburg that will stand the test of time is the recognition by governments that it is no longer they alone that do this' (Mitchell 2003). As TTPs, all three of the case studies examined here, should, in theory and in practice, have been committed to a multi-stakeholder approach, but more than this, they should have been committed to multi-stakeholder *partnership*. However, we have seen deficiencies in each of their partnership modalities, which we will now review. As we assess these partnership shortcomings, we consider whether the type of approach each TTP adopted (as discussed in the last section) affected its mechanisms and processes of partnership.

EUWI exhibited poor internal governance in its northern context for the first two years of its existence, in that the EU Member States and Commission assumed a superior status to other stakeholders, resulting in a two-tier structure and an EU stakeholder process, but not a *partnership*. In July 2004, these governance problems were addressed in a new organizational framework which better defined the roles of participants, and increased the responsibilities and involvement of non-governmental stakeholders in decision-making through the creation of a multi-stakeholder Steering Group to drive the initiative forward. At the level of partnership between the EU and Africa, the initiative was accused of being northern-dominated, as well as government-dominated. In its partnership at the country level in Africa, the Zambian case revealed a government bias, with extremely restricted participation of indigenous non-governmental stakeholders, although Zambian water and sanitation NGOs were few, and some indigenous private sector players chose not to participate. The international private sector was also notably absent, though international NGOs were well represented at the country level. EUWI has, on the whole, therefore, demonstrated weak partnership features in its different structural levels.

Accordingly, although EUWI has sought to remedy many of these partnership deficiencies, it has been characterized as a top-down, government-led venture. This hierarchical structure is related to its function, in that because it has sought to devise strategic country level and sector-wide plans to reach the MDGs, its priority has been to work at the political and governmental level within Africa, as well as at the level of harmonising donors (with an emphasis on European Member State donors). Indeed, it could be argued that for the high-level political work that EUWI chose, multi-stakeholder partnership was not necessary at all. One interviewee (2, 2003) stated that 'Partnership right now is a kind of fad, it's a kind of buzzword . . . [but] declaring that everything needs to be a partnership is not helpful.' However, donor harmonization (one of the EUWI's strategic aims), if it is to be effective, has to be inclusive, and requires, at the very least, co-operation and consultation between various stakeholders, both governmental and non-governmental. Furthermore, improved water governance necessitates non-governmental stakeholder participation at both the planning and implementation stages, though while the EU players can encourage a multi-stakeholder approach, they cannot force it upon indigent governments.

Like EUWI, PAWS struggled with its internal governance structures in its early years, which were widely seen as government-dominated. Not only did PAWS betray a government bias in its northern context, but it did so also in its African context, most notably in South Africa, where it failed to engage with actors beyond government in any substantial way. This flew in the face of its stated aims 'To encourage best practice and develop guidelines for effective and sustainable tri-sector partnering . . . [and] To demonstrate the effectiveness of multi-stakeholder engagement' (PAWS 2006a). In 2004, it tried to address these concerns through establishing new terms of reference for the initiative, which included the appointment of the PAWS Chair from outside government, and the establishment of a multi-sector Steering Group. PAWS also sought to improve partnership mechanisms through the addition of a non-governmental secondee to the Secretariat, and through holding partnership meetings outside UK government offices. PAWS, like EUWI, has also been accused of being northern-dominated, although neither initiative has gone into an African country and tried to convince it that any single approach is the correct one, or imposed conditionalities on assistance.

Again, as with EUWI, PAWS's poor record on non-governmental stakeholder participation in Africa has been affected by the type of approach it chose to take – to build capacity within existing structures. Since PAWS sought to partner with African governments first and foremost, and to address needs as identified by them, it necessarily worked within the constraints of what a partner government wanted to do, and it was up to African governments to decide whether they wanted to engage with other stakeholders. Thus PAWS's top-down government-led strategy seemed due largely to its commitment to a capacity building approach. However, recent work in Nigeria suggests that a more participatory way of working has been chosen, utilising WaterAid's contacts on the ground to give much greater emphasis to indigenous community and NGO involvement.

WAWI is different from both EUWI and PAWS in that most WAWI participants have committed funds to the initiative. This suggests that the basis for relationships between members should be more equitable than is the case in EUWI and PAWS. In practice, however, this is not the case, and the actors who have committed the most funds to the initiative (the CNHF and World Vision), and house the initiative's Secretariat (World Vision), wield much more power than do the other members. The central difference between WAWI and the other two TTPs is that although the dominant partners in all three initiatives have proven to be those providing the majority of funding, in WAWI this is two non-governmental organizations, whereas in both EUWI and PAWS it is (northern) governments.

However, WAWI did not struggle to the same degree with its internal governance structures as did EUWI and PAWS. This was partly because WAWI has fewer members, making relationships easier to manage, and partly because WAWI is less a *partnership* than an overarching programmatic framework for individual contractual funding arrangements. In general, contractual arrangements are much easier to negotiate than those required for mutual partnership. It was also partly because the more prominent roles of World Vision and the CNHF were broadly accepted by other WAWI participants at the outset, whereas the dominant roles that northern governments assumed in EUWI and PAWS were contested from the beginning. However, in time, the dominance of World Vision and the CNHF in WAWI became challenged, and tensions arose around the fact that some partners want to see the geographical areas in which WAWI works expanded beyond the boundaries of World Vision's project areas. Moreover, some WAWI partners – including USAID, WaterAid, and UNICEF – have questioned the geographical spread and largely philanthropic outlook of the CNHF and World Vision.

Like PAWS, WAWI specifically stated that it aims to 'foster a new model of partnership and institutional synergy' (CNHF 2004a), but although the CNHF has implied that African governments, as well as local communities are partners in the initiative (CNHF 2004a), in practice, their status is ambiguous, with different partners perceiving the roles of indigenous actors (both governmental and non-governmental) in different ways. This ambiguous status relates to the divergent approaches chosen by partners in WAWI, in that those partners with a philanthropic outlook who did most to shape the partnership (the CNHF and World Vision), are essentially concerned with direct service delivery and working with communities to help them build their own water systems and to empower them to be able to maintain these services after WAWI has left. On the other hand, there are partners such as WaterAid, USAID, and UNICEF who want to see a more strategic approach adopted, whereby WAWI continues with its projects, but uses them to influence indigenous government policies and implementation strategies.

The Role of Government

Role of Northern Government

In chapter two, we rehearsed two fundamental concerns about the TTP approach to the implementation of sustainable development. First, that TTPs might be used by governments to transfer what are essentially government responsibilities to other actors; and second, in the opposite direction, that TTPs might not be mutual endeavours, but dominated by governments, which engage with other stakeholders only superficially, in an attempt to avert criticism, legitimize their role, and reinforce their position. What light did the three case studies shed on these concerns?

First, in relation to the role of northern government, evidence suggests that for those partnerships where northern governments provide the majority of funding and secretarial support, the new TTP framework is not very different from the traditional government-dominated model of bi/multi-lateral development. Both EUWI (especially) and PAWS have been driven mainly by government-led pressures, reflecting their respective top-down political systems. To some extent, therefore, both EUWI and PAWS are dominated by top-down pressures, and have been used by northern governments to reinforce their positions.

On the other hand, northern government has not played as a strong role in WAWI as it has in PAWS and EUWI. We are not claiming that there was an explicit choice made by any northern government to let WAWI take on some of the functions that it should have been fulfilling, but rather that WAWI has filled gaps in provision that the US government and the governments of WAWI's focal countries have failed to meet.

Role of Indigenous Government

Second, in relation to the role of indigenous government, it was hoped that the new partnership approach to sustainable development would foster more equitable relationships between the north and south, doing away with northern-imposed prescriptions and ensuring in-country ownership. After all, as 'President Mkapa of Tanzania has said: "Development cannot be imposed. It can only be facilitated. It requires ownership, participation and empowerment, not harangues and dictates"' (Benn 2005a, iii). This section examines whether our three TTPs have empowered indigenous governments, and whether their treatment of indigenous governments has been a reflection of the approaches to development assistance employed in the unilateral/bilateral programmes which originated, respectively, in the TTPs' three northern bases of the EU, the UK and the USA.

Both EUWI and PAWS stress the role of indigenous national government in their work, whereas WAWI places less emphasis on the importance of African governmental engagement. EUWI began its work in Africa by establishing the *African-European Union Strategic Partnership on Water Affairs and Sanitation,* signed by African heads of state and the President of the European Council and President of the European Commission. At the regional level it partnered with AMCOW, and while partnering at this level was deficient in many ways, the focus undoubtedly was on partnership between the Member States of the European Union and the governments of African nations. At the country level, EUWI sought to establish country level dialogues, and the Zambian case showed they centred around national governments and associated actors, and donors – largely EU Member States, with some participation from other international donor agencies and international NGOs, as well as a limited involvement of indigenous non-governmental actors. Although the country dialogue process stalled in Zambia, partly because it was not sufficiently demand-led and was seen as externally imposed, a new strategy for country dialogues was pursued, and the Zambian process was to be reinvigorated as a result. The new strategy focuses on bringing stakeholders together in a country to create a roadmap to meet the MDGs – a process in which indigent government plays a key role as the first, and last, port of call for the EUWI.

Similarly, for PAWS, African government engagement is viewed as essential. From the outset, PAWS sought to create partnership agreements between PAWS-UK and the central governments of its three initial focal countries: South Africa, Uganda and Nigeria. In South Africa, although the emphasis was mainly on assisting local government, PAWS always worked directly with central government departments first, before working at the more local level: indeed, PAWS worked with central government mainly to identify municipalities in need of the capacity building assistance that PAWS offered. There was also a PAWS National Steering Group in South Africa, made-up of the PAWS Coordinator, and representatives from the Department of Water Affairs and Forestry, the Department of Provincial and Local Government, and the South African Local Government Association. This ensured that national government was firmly placed as a lead force in the initiative. Although in Uganda, progress was slow, in the early days PAWS did do some work with the Ugandan government on regulation. More recently, a MoU was signed with the Ugandan government's Directorate of Water Development to provide capacity building at various levels. Although PAWS also worked with the Ugandan Water and Sanitation NGO Network and the Association of Private Water Operators, this activity was undertaken under the umbrella of the MoU with the DWD. The initial work of the partnership in Nigeria turned out to be an 'abortive engagement' (Turner 2005), but an engagement with Benue State was more productive, along with similar work elsewhere in the country that has since taken place. Because of the federal system in Nigeria, the main governmental focus for PAWS was not at the central or federal level (although PAWS did have backing at this level too), but at the state level, because state governments enjoy a large degree of autonomy. This work involved non-governmental actors as well as government, but the government's role was crucial.

WAWI, on the other hand, places less stress on the role of African government in its partnership work. While the EUWI and PAWS both talk of African governments as "partners", in WAWI partnership literature, the governments of Mali, Niger, and Ghana (WAWI's three focal countries) are rarely formally recognized as "partners" in the initiative. Instead the title of "partner" is usually reserved for northern hemisphere actors (Water for the Poor 2002a; WAWI 2006). Although the WAWI website and the CNHF (Water for the Poor 2002c; CNHF 2004a) both professed that African governments and local communities were important actors in the initiative, and could even be considered as partners, interviews with WAWI partner representatives suggested that, in practice, the role of indigenous government has been patchy. While central governments are engaged to some extent, are consulted with formally, and participate in country level meetings, they remain to a large degree at the periphery of most activities, and play little active role in the partnership beyond endorsing WAWI's work. African governments certainly have no seat at the WAWI table when it comes to overall policy decisions at the headquarters level.

One reason why WAWI has marginalized indigenous governments is because its work on the ground has been fragmented. WAWI is made up of several different organizations that have co-ordinated with each other but have acted independently on their own projects under the umbrella of WAWI. These organizations hold differing opinions about the role of indigenous government, and place different emphases on the need to work closely through government structures. While many participating organizations such as USAID, Lions Club, WaterAid, Helen Keller Worldwide, the Carter Center, and UNICEF are mindful of the need to build capacity within existing structures, and thus advocate a strong role for indigenous government, World Vision and the CNHF, as the lead implementing agency and the primary funding organization respectively, are more concerned with service delivery and with having a direct impact on the lives of beneficiaries.

However, by not placing indigenous government at the heart of its efforts, WAWI runs the risk of weakening governance systems in-country, undermining those governments' roles as service providers, and confusing accountability. An NGO representative (Interviewee 1, 2003) argued that:

> the danger of TTPs is that unless they work within national budgeting and planning strategies, they may create alternate power centres . . . [and] set up different levels and types of authority and responsibility . . . most stakeholder approaches are interesting if they enhance democratic processes but don't act as alternates . . . [They need to] consciously work within the formal tiers of government because otherwise they're undermining what they're also trying to achieve which is more democratic, more accountable institutions that are responsive to people. So I think there is a danger with the stakeholder approach that it could set up a parallel system.

In the case of WAWI, it is not so much that its multi-stakeholder approach within Africa has established a parallel system. Rather, government is marginalized from its proceedings, and this may have had the same consequences as those identified by the interviewee above. In order for the TTP framework to fulfil its potential, not only must stakeholders outside government be embraced fully as partners, and a sense of mutuality achieved, but governments must play a strong role in TTPs, because of the fundamental need for political commitment, and because they are the main source of funds for development work and improved service provision.

Do these differing approaches to the role of indigenous government reflect the differing developmental policies employed by governmental donor agencies from the geographical locations of the different northern hemisphere base of each partnership? In answering this question, we must first identify the three main current approaches to donor assistance adopted by northern governments – project support; sector wide approaches (SWAPs); and general budget support (GBS). Most (if not all) donor governments undertake project work within developing countries, whereby they either implement a project themselves or fund non-governmental individuals and agencies to undertake this work, either drawn from within the particular developing country or from abroad. This approach is unlikely to disappear, but its popularity as a stand-alone developmental strategy has waned in recent years, largely because of perceptions that it leads to weak in-country ownership; often has limited impact and effectiveness; and lacks strategic focus. In response to these deficiencies, the 1990s saw the popularity of the sector-wide approach (SWAP) to development assistance grow. DFID (2006) reported that SWAPs:

> offer potential advantages over stand-alone projects, including: greater government ownership and leadership; greater alignment of donor activities with government sector policies and budgets, and greater opportunities to link sector support to national policies and poverty reduction plans; greater focus on sector-wide issues affecting performance; enhanced transparency and predictability of aid flows; enhanced donor harmonisation and reduced transaction costs; and greater opportunities for civil society engagement in sector policy and planning.

The move towards SWAPs came as 'a response to the fragmentation, and perceived limited effectiveness of aid' (Norad 2004, 4), and because there 'was a perceived need to establish and maintain forums for dialogue . . . [and that this] dialogue should lead to plans and budgets that direct the use of all donor resources' to priorities in the sector as defined by the indigenous governments (Norad 2004, 5), rather than individual donors working in an un-coordinated manner on their own projects and programmes. SWAP evaluations indicated that SWAPs were beneficial 'in respect of governments' capacity to plan and implement programmes and of donor coordination' as well as improving governments' 'understanding of problems of service delivery and access by the poor, and the translation of stated sector principles into resource allocations' (Norad 2004, 4; 13). To a greater extent than previously, aid has started to flow through government channels, rather than through channels created by donors to finance "their" projects' (Norad 2004, 4). However, SWAPs have been criticized for their tendency to set up burdensome financial management structures outside government, and 'for removing responsibility for policies on the sectors away from governments and towards a group of professionals on either the donor or the government's side of the SWAP management team' (Norad 2004, 5). Another criticism of the sector-based approach has been that funds are fungible; 'Spending donor money on [a] priority sector does nothing to guarantee that spending on [that] priority sector increases,' because a government may, as a result of donor funding, reallocate funds elsewhere that it would otherwise have spent on that sector (Norad 2004, 4).

In response to these criticisms of SWAPs, in recent years there has been growing interest in general budget support (GBS) as an alternative modality of development assistance to that of either project support or area/sector based assistance. Indeed, 'donor governments [are] now channelling \$5bn (£2.7bn) a year – about 5% of their total aid – directly to the budgets of developing countries' (Elliot 2006). 'Under GBS, rather than doing projects, donors provide cash transfers to the government which spends GBS funds on its own development programs' using its own budget systems (USAID 2004b). The argument justifying this form of aid:

> is that developing countries should be defining their own strategies for reducing poverty and for achieving the MDGs through a national political process where poverty is recognized and acted upon. Development partners should provide generic support, using Government's own procedures, to help them implement their strategies. This puts decision-making and responsibility where it belongs, with Governments, it reinforces accountability to domestic political institutions rather than to foreign donors, and it builds sustainable domestic political institutions rather than bypassing them (Norad 2004, 3).

In other words, GBS 'builds strong host-country ownership and thus it stands a greater chance of supporting sustainable development' (USAID 2004b). Furthermore, a 2006 report by the Organization for Economic Co-operation and Development argued that evidence gathered over the last few years suggested that direct budget support is 'no more prone to corruption' than funding specific projects, contrary to what some had feared (BBC News 2006). However, there were anxieties within donor agencies and sector ministries in recipient countries that the total funding certain sectors received might decrease, and that development partners might have less influence in a sector (Norad 2004, 5). Indeed, in his keynote address to a Royal Geographical Society Conference *Water for Life – the UK input*, Hilary Benn (2005b), the UK government's Minister for International Development, suggested that funding for water and sanitation had declined because both donors and recipient governments had shifted their focus to health education. This problem was compounded, according to Benn (2005b):

> because of the way we have delivered our Poverty Reduction Budget Support, particularly in Africa . . . Water and sanitation have suffered, partly because there is normally not one ministry that deals with them, but several, partly because responsibility for water may often lie at local level, and partly because we've failed to push very hard at all for Poverty Reduction Strategies to reflect the demands of the poor for better water and sanitation. As a result, only a very small proportion of our budget support has been spent on water, reflecting the low priority given to water by developing countries in their poverty reduction strategies.

Moreover, GBS 'does not always strengthen democracy or public accountability' which is better served 'by projects that support parliamentary oversight, effective media, and citizen groups' (USAID 2005, xi). Thus, as Benn (quoted in Elliot 2006), stated, GBS is not a panacea: 'Direct budget support is only one instrument . . . We have to see how it can be used in conjunction with other approaches,' and it is 'important for donors to make a full assessment of political risks before channelling funds straight to governments.' In other words, 'budget support is not always the best technique to use, but it is an important tool in directing aid' (BBC News 2006).

Nonetheless, while there are many drawbacks to both the SWAP and the GBS approach, and they are not appropriate for all aid dispersal and in all countries, what the use of these aid mechanisms by some donors indicates, is a belief in the importance of the role of indigenous government to the success and sustainability of any developmental or poverty reduction strategy. SWAPs (especially those involving budget support to a sector), and more so GBS, are moves closer to the fundamental principle that donor assistance should be demand-led, and should work to strengthen existing structures and empower recipient country governments to better meet the needs of their citizens. Both SWAPs and GBS in principle strengthen lines of accountability between indigenous citizens and their governments, instead of transferring accountability to external donors.

We can relate these three approaches to development assistance indirectly to our three case studies through the geographical locations of the partnerships. The Norwegian Agency for Development Cooperation (Norad 2004, 7) reported that 'The EU is fast becoming one of the biggest budget support donors in Africa,' and it also sustains a lot of sector budget support. Furthermore, many (although not all) of the European Union Member States (including France, Sweden, Denmark, Spain, Italy, the Netherlands, Germany, and the UK) employ the GBS method of support. The UK is one of the strongest proponents of GBS, spending 25 per cent of its aid in this way (Elliot 2006). 'DFID has made a clear policy commitment to use general budget support in . . . cooperating countries that meet a set of economic, governance and other criteria' (Norad 2004, 6). With regard to SWAPs, many of these same countries are also engaged in SWAPs. Although DFID (2006) conceded that its aid assistance:

> can take any form – budget support, projects, technical cooperation, [or] policy dialogue. DFID believes that donors should commit to progressive reliance on government procedures to disburse and account for funds (helping governments to strengthen . . . disbursement and accounting procedures as necessary) . . . DFID is committed to . . . putting more of its aid through government systems where conditions are right. SWAPs are a good way of doing this, particularly when embedded in a national poverty reduction strategy. Budget support is DFID's preferred means for transferring financial aid within a SWAP, combined with policy dialogue and technical cooperation, where appropriate.

Indeed, in March 2005, Hilary Benn announced a change in the way UK aid was delivered in a new policy paper, *Partnerships for Poverty Reduction: Rethinking Conditionality*, whereby it would no longer 'make its aid conditional on specific policy decisions by partner governments or attempt to impose policy choices on them . . . Instead, the UK Government will agree benchmarks with partner countries which look at the impact of the decisions they take on reducing poverty and improvements in health and education' (DFID 2005).

On the other hand, 'Countries such as the United States and Japan tend to be less keen on providing direct budget support, fearing that money will be squandered unless there is accountability to donors' (Elliot 2006). Similarly, in terms of SWAPs, although 'There is broad international endorsement for SWAPs . . . donors differ on which instruments they prefer to use within a SWAP (for example the USA and Japan prefer project aid over budget support)' (DFID 2006). As we learned in chapter three, in Zambia, while USAID attended donor harmonization meetings, and in-country staff tried to align themselves to sector-wide processes, constraints from within America were placed on the agency staff, which meant that they could not formally sign up to any harmonization processes. Indeed, 'Most USAID assistance goes through projects . . . [whereby USAID] fund development projects and manage project implementation, using NGOs or contractors,' and USAID also provide technical assistance and training to governments (USAID 2004b). Thus, while USAID are happy to try and align their project work with other donors to contribute towards an overall sector strategy devised by the indigenous governments, they are reluctant to give aid directly to the government in generic terms, or even on a sector basis. The reason for this stance is, as USAID readily admit, that its mission is underpinned by the goal of furthering America's foreign policy interests and benefiting American businesses. Accordingly, in exchange for aid, foreign nations often have to purchase American services or goods, and the majority of USAID contracts go to American organizations, public or private, ensuring that a large proportion of foreign aid returns to America. Indeed, in a report from the *Reality of Aid* project (a north/south non-governmental initiative analysing poverty reduction and development assistance) in 2000 (quoted in Shah 2006), it was suggested that '71.6% of its bilateral aid commitments were tied to the purchase of goods and services from the US.'

In considering the direct relevance of this discussion to our three case study partnerships, clearly none of the case studies reflect the principles of GBS, because they are all sector/area based initiatives, with WAWI being also largely project/service delivery-based. Moreover, while these partnerships have worked within SWAP frameworks, they have certainly not engaged in any kind of water and sanitation budget support, since EUWI and PAWS have no funding available to disburse in this way, and WAWI concentrates on its own project work. However, it is interesting to note that the EU and many of its Member States, especially the UK, are proponents of both SWAPs (especially budget support SWAPs) and GBS approaches, which points to their conviction that indigenous government is of paramount importance and should be at the centre of any aid programme. USAID, however, while it has made some effort to work within SWAP programmes, placing indigenous government at the fore, rarely engages in budget support or "basket funding" of a sector, suggesting a reluctance to hand the reins over to African governments and relinquish control on the spending of its aid.

It is, therefore, noteworthy that both EUWI and PAWS (based in Europe and the UK respectively) have placed great emphasis on the role of African governments (even if in practice the partnering mechanisms and modalities between northern partners and their African counterparts have been deficient – especially in the case of the EUWI), whereas within WAWI, the role of African governments has been more ambiguous, with some participants referring to them as partners, and others preferring to regard them as stakeholders. In other words, EUWI and PAWS are much more demand-led than is WAWI. Therefore, the ODA strategies and outlooks assumed by the bilateral development programmes of the donors in the northern bases of the three partnerships go some way to explaining the differences in approach taken to indigenous government by the partnerships.

However, while the UK government plays a strong role in PAWS, and the EU Member States and European Commission play a strong role in EUWI, USAID, while a partner in WAWI, is not the principal funder or implementer, and does not host the Secretariat, as governments (or staff seconded from government agencies in the case of the EUWI) have done in the two other partnerships (although the PAWS Secretariat recently moved to be hosted outside of government). The US government, therefore, has not had such an influential role in the initiative as other northern governments have in EUWI and PAWS. Indeed, the most influential organizations within WAWI are World Vision and the CNHF, in that their outlooks have done most to shape the initiative. Nevertheless, it is likely that the reason why USAID joined WAWI was that it was project-based with a high-level of external control, and USAID primarily fund American organizations to undertake project work.

It would seem, therefore, that the organizations that have exerted the most influence over a partnership have been those that have committed the most funding and provided the Secretariat for the initiative. In the case of EUWI, this was the European Commission and the EU Member States; for PAWS it was the UK government, most notably DEFRA; whereas for WAWI it was the CNHF and World Vision. In EUWI and PAWS, existing approaches to governmental bilateral aid are therefore likely to have had an impact on partnership operations, while for WAWI this is less likely. The fact that USAID were involved in WAWI may have played a small part in shaping its programme, but the CHNF and World Vision have had the most impact, although USAID would not have signed-up to WAWI if the working methods of World Vision and the CHNF were discordant with its own vision.

The fact that in WAWI, northern governments, or a northern government is not the lead partner, also explains why this initiative was not originally signed up as a TTP. Being led by the CNHF and World Vision, it was not imperative that it was celebrated at the international level as a part of a governmental strategy to meet the MDGs, as PAWS and the EUWI were. It was USAID who trumpeted WAWI's existence at the international level, but for many years WAWI remained un-registered as a TTP, largely because this was not something of great importance to the founding members and largest funders of the initiative.

Evaluating Partnerships through Outcomes Achieved

The above analysis, and indeed most of the analysis in this book, has concentrated on approaches to partnership, partnership processes, and their ability to fulfil the potential that this mode of operation offers. These process issues of partnerships, as chapter two established, are central to their evaluation. However, as was also noted previously, TTPs were not created as an end in themselves, they were created as a means to achieve certain goals (Haque 2004, 279–280; Torjman 1999, 16). Thus, we now undertake a comparative analysis of the outcomes of these partnerships and their tangible achievements. However, before doing so, we should briefly consider the monitoring systems and reporting mechanisms utilized by the three partnerships, since these form much of the basis of our information on partnership outputs.

All three partnerships have struggled with monitoring issues. PAWS has experienced difficulty defining and implementing monitoring tools, but it does operate in a largely transparent manner and regularly reports on its progress to the CSD Secretariat and to the public at large. Like PAWS, EUWI has struggled to find the right monitoring mechanisms, but unlike PAWS, it has not been as proficient at sharing information on its progress with the CSD. It does, however have a website that provides up-to-date information on the initiative, and it holds an open multi-stakeholder forum every year to be accountable to the public at large. For its part, WAWI reports its progress to the CSD Secretariat, but in terms of wider dissemination of information for public scrutiny its operations remain largely behind closed doors with a poorly maintained website and with almost all its partnership reports and policy documents seen only by its members.

To facilitate an evaluation of the concrete achievements of the three partnerships, we return to the nine criteria established in chapter two for the assessment of partnership outcomes:

1) Realize objectives that no single actor could achieve alone and 'not merely reflect existing arrangements' (UNDSD 2003).
2) Be linked to global and national goals and strategies in order that their contribution is meaningful.
3) Avoid duplication of activities and enhance coordination between various stakeholders.
4) Provide innovative solutions combining the resources, capacities and knowledge of all involved.
5) Empower recipients and 'where relevant, result in . . . capacity building in . . . developing countries,' enhancing the effectiveness and sustainability of project outcomes through local ownership (UNDSD 2003).
6) Mobilize new resources over and above those already being supplied by governments.
7) Complement but not replace governmental efforts.
8) Avoid undermining the accountability mechanisms between citizens and their governments.
9) Meet the partnership's own espoused objectives.

EUWI aspires to transcend the constraints of existing bilateral aid structures, at the European level, the regional level, and the country level, but it is unclear whether it has met this aspiration. Its work is aimed at linking to national, as well as to global, goals and strategies, as well as enhancing donor co-ordination to avoid duplication of in-country efforts. At the European level, its governance structures improved over time, and synergy has been evident, but at the regional and country levels its progress is less palpable. The Zambian case study demonstrated that EUWI failed to find innovative solutions to Zambia's water and sanitation problems; it did not succeed in empowering recipients or in achieving in-country ownership; and no new resources were mobilized. EUWI has not, however, been used by governments as a tool to replace their own bilateral aid efforts (although some EU Member States did refer to EUWI when asked what their country was doing to help achieve the water and sanitation MDGs), and accountability mechanisms between citizens and their governments have not been undermined. But EUWI has failed to achieve any concrete outcomes. As one interviewee (6, 2004) remarked, 'at the end of the day, when we reach 2015 nobody will be able to say that this borehole is a result of the European Union Water Initiative, but . . . without this kind of process to improve what everybody is trying to do we have less chance of succeeding.'

Like EUWI, PAWS sought to develop and improve its governance structures in the UK, enhancing its synergistic properties. But it is on the ground in Africa where synergy matters most, and in South Africa, most PAWS work was undertaken by the UK in collaboration with the South African government (at national and local levels) and the UK private sector, but with virtually no engagement from the UK-based NGOs or indigenous non-governmental stakeholders. However, work in Nigeria indicates that synergy between the three sectors from the UK and between various indigenous stakeholders has improved, and PAWS has provided innovative solutions to combine the resources, capacities and knowledge of all those involved. PAWS has always been careful to work with the national governments of the countries with which it partnered, to avoid duplication of activities and to create in-country ownership, and its focus has consistently been on capacity building. Although PAWS, for the most part, has failed to find new sources of funding beyond those of the UK government, its programme in Benue State did include funds from WaterAid Nigeria (although this would probably have been spent in Nigeria regardless of the PAWS initiative), and in-kind contributions have played a large role in PAWS's work, most notably from the UK private sector, which constitutes a new type of resource. PAWS has not replaced governmental efforts but has complemented them, and it has not undermined the public accountability of African governments. As PAWS concentrates on capacity building, its achievements are harder to gauge than if it were concerned with service delivery. Like EUWI, PAWS's performance in delivery of water and sanitation services has not been as successful as was hoped, and its practical impact has been limited. However, it has gone some way to achieving some of its objectives, and where it has failed, it has drawn lessons from its failure.

With regard to WAWI, it is unclear whether it has achieved any significant degree of synergy, added value or holistic solutions. On the whole, it has operated as a series of projects in a given geographical area, with individual grant agreements between certain partners, coordinated under the overarching banner of WAWI. In these projects, it has increased water and sanitation provision to people who were un-served or under-served, and it has achieved some level of recipient empowerment through its work with communities in building their capacity. However, in-country ownership would have been better promoted if WAWI had worked within existing governmental structures. While WAWI's work has complemented, rather than replaced, governmental efforts, by not actively engaging with African governments and placing them at the centre of the partnership's strategy, it has run the risk of setting up a parallel system of accountability. However, on a more positive note, WAWI does not depend on government funding; instead it draws together the resources of many organizations. Moreover, as an initiative which has focused on direct service provision, WAWI has been able to quantify many of its successes (something which EUWI and PAWS are unable to do with their focus on sector-wide governance and capacity building respectively). The figures produced show that WAWI has performed well in terms of the numbers of people who have gained access to services as a result of its work.

Conclusion

Of the three approaches pursued by the three partnerships examined in this book – first, sector-wide governance; second, capacity building within existing government structures; and third direct delivery of services – EUWI concerns itself mainly with the first; PAWS focuses on the second, but with a commitment to improving governance, and with a nod towards service delivery; and WAWI centres around the third, with some capacity building (although largely outside existing government structures) and a recent interest in sector-wide governance. In evaluating their performance, we assessed their processes and their outcomes.

With regard to their processes, the main test was how genuinely they have worked as partnerships. There are two questions in this test: first, are the TTPs too northern-dominated? Second, are they too government-dominated? On the first question, EUWI and PAWS are less northern-dominated than is WAWI, because they place indigenous government at the heart of their initiatives (at least in theory), whereas WAWI does not conceive such a significant role for African governments in its implementation work. However, this commitment to indigenous government means that EUWI and PAWS are less open to non-governmental stakeholder participation (most significantly within Africa), leading both to be accused of being government-dominated (although PAWS less so than EUWI). WAWI, on the other hand, works very closely with communities, although not in a true sense of partnership.

The fear that the TTP framework would be used by governments to abdicate their responsibilities is unfounded. Both EUWI and PAWS have proved to be strongly driven by government-led pressures, and even WAWI shows no signs of being used as an excuse for inaction by government. However, in EUWI and PAWS, government domination has threatened to undermine the formation of genuine partnership relationships, while in WAWI, government has been too weakly engaged to ensure the sustainability of its work. Achieving the optimal level of government involvement is clearly a difficult balancing act, but one that cannot be avoided and must be tackled head-on if the promise of the TTP framework is to be achieved. For Simpungwe *et al* (2007, 191), governments themselves bear a heavy responsibility to get this balance right: 'The State . . . has a crucial role in developing new capacities to empower (*all*) stakeholders to overcome their "*limitations*" and become legitimate owners of the process.'

In terms of outcomes, one may, at first glance, deem WAWI to have been the most successful, since it alone seems to be on target to meet is own service delivery goals. However, we cannot yet judge how sustainable these outcomes will be. Moreover, while EUWI has failed to meet any outcome goals, and PAWS has a patchy record of valuable work, many of their objectives are more complex, and much harder to achieve and quantify than the direct provision of water and sanitation services. This is not to detract from WAWI's achievements, which are considerable, but WAWI itself has an objective which is difficult to quantify: to foster a new model of partnership replicable elsewhere (CNHF 2004a). It is with this aim that WAWI wrestles most, and both PAWS and EUWI struggle with it too (even if EUWI has not explicitly stated that creating a model for multi-stakeholder partnership is one of its aims).

In short, the evidence from these case studies suggests that the full potential of the multi-stakeholder partnership approach has not yet been realized. Within Africa, the ideal of multi-stakeholder partnership has not been achieved in any of the three TTPs studied, but has been confined largely to northern hemisphere operations (and even here partnership between different stakeholders has not always been equitable). However, efforts to work in a partnership manner have improved over time in all three cases, both in the north and within Africa. In our final chapter, we summarize the findings of this study, and explore the wider implications of these findings in relation to the theoretical framework of sustainable development established in chapter two.

7

CONCLUSION: SUMMARY AND WIDER IMPLICATIONS – THE ROLE OF TTPS IN GOVERNANCE FOR SUSTAINABLE DEVELOPMENT

Summary of Findings

The inception of the TTP framework was an innovation in the improvement of governance for sustainable development both because of its multi-stakeholder character, and because of its conceived nature of the proper relationship between stakeholders, that of partnership. The case studies examined in this book have all made contributions to improved governance in their different ways, even though their methods have been flawed, and their partnership modalities deficient. There are strong indications from the case studies that inclusive and participatory ways of working do have significant benefits compared with governmental/top-down prescription for the implementation of sustainable development, and that multi-stakeholder partnership working can achieve much more than individuals or organizations working alone.

With regard to partnership *processes,* the three TTPs have attempted to forge equitable relationships between all participants (although WAWI less so than EUWI and PAWS), even if they have not always achieved this. However, a trend displayed by all three partnerships is that those organizations which have brought the greatest financial resources to a partnership have usually possessed more influence than other partners over partnership activities, even if they have chosen not to exploit this power. Moreover, EUWI and PAWS (although EUWI more so than PAWS) have both been criticized for over-emphasising the role of governments, and for not sufficiently engaging non-governmental stakeholders in their activities, in both their northern and southern contexts, but especially within Africa, while WAWI has been criticized for not assigning sufficient importance to the role of indigenous government, thereby weakening governance in the countries where it has been working, by damaging lines of accountability between citizens and their elected representatives. It is clear that the role of governments in TTPs is of fundamental importance to the TTP framework, and is something that has to be addressed by all TTP initiatives when they first begin to develop their strategies, because a careful balance needs to be established and maintained between the opening up of the governance of public affairs to non-governmental stakeholders, and the preservation of governmental responsibility and accountability.

With regard to partnership *outcomes,* the three case studies observed in this book have, on the whole, demonstrated an aim to be demand-driven. But concrete achievements have been patchy: EUWI has so far proved unsuccessful at meeting any of its concrete goals; PAWS's impact has been limited and selective; and although WAWI has had reasonable success in meeting many of its goals, it is unclear how sustainable these outcomes will be.

The evidence suggests, therefore, that in terms of both processes and outputs, the full potential of the TTP approach has yet to be achieved. However, all three partnerships have shown reflexive tendencies and an appetite for adaptation and organizational learning (though EUWI and PAWS more so than WAWI). EUWI learned from the mistakes it made in terms of both its internal governance structures and its approach to work on-the-ground in Africa, and it reoriented its approach accordingly. Similarly, PAWS adapted to improve its internal processes and to exert a greater external impact, drawing on previous failings to guide its future trajectory. Finally, WAWI realized the value of a strategic approach and the need to create more synergy amongst partners, and adjusted its techniques accordingly. What this shows is that TTPs involve learning processes, and learning takes time. However, if the lessons already learned by existing TTPs could be disseminated beyond the individual partnerships, then future initiatives would have more chance of speedier results. This leads us to consider the wider implications of the findings of this study, and in the rest of this chapter we discuss the place of TTPs in global governance for sustainable development, focusing on two cardinal issues – voluntarism and accountability.

Can Voluntary TTPs Replace Binding Commitments?

In answering this central question, it is instructive to compare TTPs with the UN's Global Compact. The UN is no stranger to controversy over partnership arrangements and frameworks, as the Global Compact indicates. In a speech made to the World Economic Forum in January 1999, UN Secretary General Kofi Annan, 'challenged business leaders to join an international initiative – the Global Compact – that would bring companies together with UN agencies, labour and civil society to support universal environmental and social principles' (UN Global Compact 2006a). The operational phase of the Compact was launched in July 2000 at the UN headquarters in New York, and it now includes more than 2,500 companies: 'from all regions of the world, international labour and civil society organizations are engaged . . . working to advance ten universal principles in the areas of human rights, labour, the environment and anti-corruption' (UN Global Compact 2006a). The Compact is, therefore, a multi-stakeholder initiative focusing on enhancing dialogue and collaboration to promote new forms of "good governance" (Utting 2002). The initiative is purely voluntary and:

> is not a regulatory instrument – it does not 'police', enforce or measure the behaviour or actions of companies. Rather, the Global Compact relies on public accountability, transparency and the enlightened self-interest of companies, labour and civil society to initiate and share substantive action in pursuing the principles upon which the Global Compact is based (UN Global Compact 2006a).

In other words, the initiative focuses on encouraging businesses to adopt socially responsible and environmentally sustainable policies, and report on them, but it is not a regulatory instrument: 'it is not designed, nor does it have the mandate or resources, to monitor or measure participants' performance' (UN Global Compact 2006b).

Utting (2003) claimed that proponents of the Global Compact initiative saw it as 'an innovative and pragmatic approach that can reform corporate culture by instilling new values and mobilise the resources of big business for social and sustainable development . . . [It] is regarded as an exemplary form of "good governance", where cooperation and voluntary approaches win out over conflict and heavy-handed regulation,' and which promotes social or organizational learning 'where business and other stakeholders learn through multi-stakeholder dialogue, analysis and networking.' However, the Compact has been at the heart of a great deal of controversy since its inception. Some of its most notable and vocal critics in the first few years were a grouping of NGOs under the banner of The Alliance for a Corporate-Free UN, led by the US-based CorpWatch. These NGOs claimed that many corporations which had signed up to the Global Compact, such as Norsk Hydro, Aventis, Nike and Rio Tinto, were still violating its principles in many areas of their businesses. In other words, corporations were using the initiative as a smokescreen, while continuing to engage in practices which contravened the principles that made up the backbone of the agreement. Thus, as Utting (2003) noted, critics of the Compact were:

> concerned that it may be doing more to enhance the reputation of big business than aiding the environment and people in need. They are worried that companies with a reputation for malpractice have been welcomed into Global Compact, and that the conditions imposed on business to comply with the principles are very weak . . . .[with] no monitoring of compliance. The focus on best practices diverts attention from malpractice, 'greenwash' and structural and other factors that encourage corporate irresponsibility or a 'business-as-usual' attitude.

Of course, the TTP framework differs from the Global Compact in that it is not about encouraging businesses to run their organizations in more socially and environmentally responsible ways (although TTPs do encourage the private sector to be involved in global solutions to social problems), and it does not entail direct engagement with the UN for each partnership participant (although the UN is a partner in many TTPs, and at least one organization from each TTP has to engage with the UN to register their partnership with the CSD). Nonetheless, some criticisms of the Global Compact also apply to TTPs, including the charge that such voluntary initiatives help organizations to avoid compulsory regulation. As Utting (2002) observed of the Global Compact, there was a fear that its creation reflected 'a shift in approach' to UN-transnational corporation [TNC] relationships:

> whereby lukewarm voluntary initiatives have crowded out important mechanisms and institutional arrangements involving new forms of international law, oversight or monitoring of TNC activities, mediation or arbitration of disputes, and critical research into regulatory alternatives, as well as on the social, environmental and developmental impacts of TNCs.

A similar criticism has been made of the TTP framework, that it diverts attention from the need to impose international obligations on states to provide water and sanitation services to their populations. Some of the other sectors discussed at the WSSD under the WEHAB (Water, Energy, Health, Agriculture, and Biodiversity) agenda already had binding international treaties dealing with them. For example, for energy there was the Kyoto Protocol, and for biodiversity there was the Convention on Biological Diversity, and within that, the Biosafety Protocol. In the case of water and sanitation services, however, there was, and still is, no such treaty, although there have been calls for the adoption of an accord which recognizes and protects the human right to water (since the UN Committee on Economic, Social and Cultural Rights affirmed in 2002 that access to adequate safe water was a fundamental human right).

However, some governments are strongly opposed to legally-binding commitments, most notably the United States, which has not ratified either the Kyoto Protocol or the Convention on Biological Diversity. Significantly, the US government has been among the most vocal supporters of the TTP approach: Jonathon Margolis (2004) of the US State Department noted that Secretary of State Colin Powell asserted that 'partnership is the watchword of U.S. strategy in this administration.' Indeed, in the run-up to the WSSD, Friends of the Earth (2002) argued that:

> The Bush Administration has been a strong proponent of Partnerships throughout the preparatory process. The negotiation

> tactics of the United States during Prepcom 4 in Bali, Indonesia [one of the preparatory meetings for the WSSD] confirmed that the administration see these partnerships as an alternative rather than a complement to binding intergovernmental commitments.

Unsurprisingly, therefore, the US government's backing of TTPs reinforced some commentators' anxieties that the TTP approach is being trumpeted by governments who wish to avoid obligatory commitments to action: 'What is new and worrying with the Johannesburg Type 1/ Type 2 dynamic is it appears that governments are on the verge of abdicating their own responsibilities to deliver on their political commitments to voluntary initiatives' (Friends of the Earth 2002).

Defending TTPs against this charge, Margolis (2004) argued that what the inception of the TTP framework did was, first, shift the emphasis away from protracted negotiations towards implementation; and second, increase stakeholder participation in CSD decision-making:

> UN meetings would now be results-oriented. Partnerships are about delivering results. They are measured by whether they can provide a direct impact on the ground. With partnerships on the table, delegations were forced to think hard not just about how to get through the negotiating process but also what they were going to bring to UN meetings. This marks a sea change in how our delegations and others approach UN meetings . . . [Furthermore] partnerships made it even more essential that those actually carrying out the work participate in the discussions. Our results-oriented discussions required the active engagement of all stakeholders, not just governments.

However, although there is broad agreement that 'business-as-usual and strictly government-based dialogue is not the way to go, and we will not deliver on the targets if we do not change our way of working' (Interviewee 4, 2004), there is a wide divergence of opinion over what role TTPs could play in meeting the water and sanitation MDGs, ranging from those who say that the goals cannot be met without TTPs (Interviewee 23, 2005); to those who say that TTPs will have limited effect on the ground (Interviewee 3, 2003). Our view is that while TTPs provide great potential for change and improvement, the scale of the problem is so great that it cannot be met through partnerships alone, because 'you need huge infrastructure extension and huge investment' (Interviewee 8, 2004). Indeed, 'The UN estimates that in order to meet the Millennium Development Goal for water, 270,000 new connections will have to be made each day, and more than twice as many to meet the goal for sanitation' (UN 2004), and although estimates on the financial requirements to meet the MDGs on water and sanitation have varied, one popular estimate is that it will 'require between US$ 14 billion and $30 billion a year on top of the roughly US$ 30 billion a year already being spent' (Mwanza 2003, 109).

Such levels of investment have to come largely from governments, not least because the private sector is (on the whole) not prepared to commit substantial funds to develop infrastructure in the developing world because these operations are too risk intensive. It was thought that partnerships would be able to harness funds beyond those provided by governments (and more specifically those of the private sector), but as the CSD Secretariat reported in 2006 (UNESC 2006a, 13): of the 319 registered partnerships, 256 had disclosed information on the source of their funding and, of these:

> 75 per cent reported having funding from Governments, 35 per cent are receiving funding from intergovernmental organizations (including United Nations system organizations) and 20 per cent are receiving funding from private sector donors . . . In addition to that, 16 per cent of partnerships have reported receiving grants from NGOs, while 6 per cent are receiving funding from other sources, such as foundations.

Thus, governments remain the main providers of financial resources – the voluntary initiative cannot replace, but only complement, the work of governments.

The Accountability Shortfall

The second cardinal issue raised by TTPs is their accountability. Indeed, accountability is a generic problem for partnerships (Brinkerhoff 2007, 74–76; Meadowcroft 2007, 197; Biermann *et al.* 2007a, 247–248; Biermann *et el.* 2007b, 290–292; Hale & Mauzerall 2004, 226–229).

An important commonality between the Global Compact and the TTP framework is that both frameworks have been accused of providing weak accountability mechanisms. The problem is that organizations involved in the Compact and the TTPs have gained legitimacy through joining up to these initiatives and thus appear to have UN endorsement, yet there are no substantial overarching monitoring, evaluation, or regulation structures in place for either framework. It is true that in 2005, a limited attempt was made to remedy the accountability shortfall in the Compact, as a result of which, a number of changes were implemented 'to safeguard the Global Compact's integrity' (Global Policy Forum 2005):

> Restrictions on the use of the UN and Global Compact logos have been made more explicit . . . Moreover . . . participating companies that do not communicate their progress for two years in a row will be regarded as 'inactive' and will be so identified publicly on the Global Compact website. Finally, the prior informal system for reviewing complaints in the event that a participating company may be engaged in systematic and egregious actions that undermine its commitment to, and the reputation of, the Global Compact has been made more detailed and transparent. In each instance, the first aim will be to find ways whereby the company in question can resolve the matter at hand, but the GCO [Global Compact Office] reserves the right ultimately to revoke participant status . . .

However, no such attempt has been made by the TTP framework. Instead of assuming the role of overseer, the UN has essentially allowed TTPs to be self-regulating – it has provided a model and left it up to other actors to take the process forward. There are only three main roles for the UN with respect to TTPs (other than being a partner itself in some initiatives): it is responsible for the continuing registration of TTPs on, and the updating of, the CSD Partnerships Database, launched in February 2004; it convenes the CSD Partnership Fair (which takes place at the annual meetings of the CSD); and it produces biennial reports on the partnerships. The Database includes all partnerships registered with the CSD, information about them being based on voluntary self-reports. The Partnership Fairs 'provide a venue for showcasing progress in existing Partnerships for Sustainable Development, launching new partnerships and networking among existing and potential partners,' as well as offering an arena for participants to 'learn from each other's experiences' (UNDSD 2006c). The CSD Secretariat's biennial reports are produced for the CSD's biennial review sessions, compiled from the information the CSD Secretariat receives from registered partnerships. These reports serve 'as a basis for the Commission on Sustainable Development's discussion . . . on the contribution of partnerships to the implementation of the intergovernmentally agreed sustainable development goals and objectives' (UNESC 2004, 1).

Although none of these three activities amount to the systematic monitoring or calling to account of TTPs, many observers approve of the present arrangement. For example, one WAWI participant (Interviewee 24, 2004) suggested that 'it's good to have . . . an umbrella that you can go to which houses all the different partnerships so you can learn about them . . . [but] I don't necessarily want the UN to take over and ensure that we're running it effectively.' Similarly, another interviewee (4, 2004) asserted that 'the role of the UN is really to foster the multi-stakeholder dimension and approach . . . [and] also to keep the political momentum' going, not to monitor and enforce the framework. However, other critics have argued that the UN, and in this case the CSD Secretariat, has not gone far enough in its approach to TTPs to ensure that they comply with the guiding principles designed for them. Glasbergen (2007, 18) claimed that 'the UN has relegated itself to . . . endorsing any partnership that pertains to sustainable development.' As an African Civil Society Network on Water and Sanitation (ANEW) representative (Interviewee 9, 2004) noted at CSD-12:

> the problem with these partnerships is they are being initiated but they are not necessarily reported on . . . most of them have not sent back their reports . . . so what this means is that these partnerships can only be poorly analysed by governments, poorly analysed by civil society, and not understood by many . . . there are no checks and balances . . . [and] there is a lack of accountability . . . How do they contribute to the Millennium Development Goals in tangible terms? . . . how are they improving people's lives?

Nicola Clarke (2005) (a member of the PAWS Secretariat when interviewed) maintained that although being registered with the CSD 'gives us credibility so that we are able to go to a country and say "we are a UN registered partnership",' this credibility can easily be abused, and many of the partnerships are failing to keep the information on the CSD Database up-to-date because they are not obliged to do so. When, in January 2003, the CSD Secretariat requested that all registered partnerships update the Secretariat on their status, they only received responses from half the registered partnerships (UNESC 2004, 4). In November 2003, lead partners from all registered partnerships were again requested by the CSD Secretariat to update their partnership database entries, but by 31 January 2004, only one-third of all partnerships had responded to this request (UNESC 2004, 4). Although, in total, nearly two-thirds of all registered partnerships responded to one or both of the update requests in 2003, this meant that over a third of all partnerships did not respond to either request (UNESC 2004, 4). In 2006, the CSD Secretariat reported that 'Requests for updates are sent out annually . . . [but] getting updated information remains a challenge' (UNESC 2006a, 19).

The CSD does not even know whether all the TTPs registered with it are still operating. Clarke (2005) claimed that 'you could probably knock a good third of the partnerships off the list because they've failed already.' This reflects Mitchell's (2003) point that 'partnerships that are formed just to get registered so that people have got things to talk about at big meetings, never last.' Saywell (2004) of the Water Supply and Sanitation Collaborative Council urged that 'There needs to be some sort of quality control and quality assurance' to identify which of the hundreds of partnerships registered with the CSD were still functional.

This is not to deny the value of the data on TTPs that the CSD has produced, including summaries in its biennial reports on issues such as partnership numbers, sectoral coverage, geographic coverage, time frames, funding and resources, implementation mechanisms and progress since initiation, even though these reports have been brief and limited largely to statistical analysis. Nor is it to deny the value of the CSD Partnership Fairs as learning foras: for example, at CSD-14 in 2006 there were 28 partnership presentations, 25 partnership information desks, and 7 interactive discussion sessions (UNESC 2006b, 49). The CSD Secretariat has also been trying to build an online knowledge network (UNESC 2006a, 20): an online forum for registered partnerships has now been established 'to help foster alliances, share lessons learned, encourage the sharing of best practices as well as follow up on meetings, conferences, and publications,' and this could prove useful if partnerships choose to utilize it (UNDSD 2008c). The complaint is not that these data gathering activities are valueless, rather that they do not go far enough; in particular, the CSD has failed to undertake any kind of evaluative analysis (Bruch & Pendergrass 2003). As Clarke (2005) put it:

> PAWS and other partnerships in the UK have been suggesting that the CSD Secretariat should be providing some kind of gap analysis focusing on how partnerships are progressing. Are partnerships progressing? . . . There are lots of bits of work going on around evaluating partnerships . . . but somebody needs to pool all this together and give us a central place to go to see how partnerships are working and if they are working . . . it is difficult to compare partnerships because they are all so different but that is a function the CSD Secretariat should be doing more of . . . So far they have only produced reports at CSD that say we should be doing more on partnerships because they're valuable, but they don't say why they are valuable and what lessons have been learnt. A central database of information about how partnerships have progressed and which problems caused them to fail, or how partnerships evolved or overcame problems, would be useful to any other partnership looking for guidance . . . and the CSD Secretariat would be ideally placed to co-ordinate this.

An interviewee (5, 2004) argued that the UN itself did not necessarily 'need to monitor partnerships . . . but . . . there should be a mechanism where people show that they have achieved something and done something,' though this interviewee (5, 2004) cautioned that 'the dynamics would be destroyed' if the TTP framework 'became too centralized,' so a careful 'balance needs to be achieved.' The UN's apparent reluctance to closely scrutinize the TTPs has been explained by some observers as caused by its fear of offending northern governments by being drawn into investigating how far the TTPs merely rubberstamp international inequalities between the north and the south. An alternative explanation is that the UN has not had either the resources or the mandate to take on the role of monitoring all TTPs.

Nevertheless, the 'CSD are not fulfilling their potential in terms of what they could be doing with partnerships' (Clarke 2005). At the very least, some integrity measures such as those employed in the Global Compact would be helpful. Backstrand (2006, 300) refers to this as 'reputational accountability', which entails a process of naming and shaming. As one consultant and NGO representative in the EUWI (Interviewee 10, 2004) asserted of the TTP framework, 'If you're going to have something that you've set up as a club then there have to be some rules for the club otherwise anyone can join and spoil the party.' Furthermore, Morley (2004) of Freshwater Action Network, suggested that what was needed from the UN was 'a little more guidance on accountability principles, stakeholder participation, [and] on good quality principles of partnership, but it's not really happened.' Some kind of partnership evaluation and analysis would, at the very least, help to pool lessons which have been learnt by TTPs so far. As interviewee 14, (2004) urged, 'we've got to very quickly promulgate the successes and failures [of different TTPs] . . . so that others aren't making the same mistakes . . . There's got to be a best route for working in certain environments;' that kind of information has to 'feed into the UN and then be passed out somehow.'

Alongside a resource providing information on lessons learnt on TTPs, it would also be extremely valuable if CSD could put in place more mechanisms to support existing partnerships and help them overcome specific problems such as creating business plans and building effective governance structures.

Overall Conclusion

The transformative power of multi-stakeholder partnership is not just a pipe dream; the case studies investigated in this study have proved that, while they are flawed in living up to the ideal principles of partnership, and deficient in fulfilling their theoretical potential, they could play an important role in governance for sustainable development, in terms of, first, the achievement of internationally agreed objectives, and second, the institutionalization of the role of non-governmental stakeholders in the implementation of sustainable development. Indeed, the very existence of TTPs is proof of a positive shift in the way governance for sustainable development is conceived.

It is clear that some of the criticisms levelled at the TTP framework are justified, and have to be addressed in the coming years. However, while the TTP framework and the initiatives encompassed within it may not be flawless, they demonstrate a willingness to change the way sustainable development is addressed at the international level. The old institutional structures, dominated by governments, are being opened up to other actors, thereby acknowledging the value of non-governmental stakeholder engagement in sustainable development decision-making and its implementation. However, this change should not be accompanied by an abandonment of ultimate governmental responsibility: the balancing act between governmental leadership and genuine participatory non-stakeholder engagement lies at the heart of the long-term success of the TTP framework.

While partnership working offers great potential, it is not a panacea, and it is not appropriate in every situation. Indeed, it is not clear that all of the undertakings by the initiatives discussed in this study required to be completed in "partnership", although the multi-stakeholder nature of their endeavours should certainly be encouraged. Moreover, as we have learnt, partnership cannot be imposed. Nevertheless, the TTP framework may be the first of many permanent and valuable shifts towards a more inclusive and equitable way of managing economic, social, and environmental concerns, under the rubric of sustainable development.

REFERENCES

A-EUWI [Africa-European Union Water Initiative] (2005) *Summary and Lessons Learned from the Initial Steps for Starting the Country Dialogue in Zambia,* January 2005, Lusaka: Africa-EU Strategic Water Partnership on Water Affairs and Sanitation

A-EUWI [Africa-European Union Water Initiative] Zambia (2005a) *Minutes of Meeting: A-EUWI at the European Commission in Lusaka, Zambia*, held on 26.01.2005, copy provided by R. Werchota

A-EUWI [Africa-European Union Water Initiative] Zambia (2005b) *Minutes of Meeting: A-EUWI, Zambian Country Dialogue at the Ministry of Finance and National Planning in Lusaka, Zambia,* 16.02.2005, copy provided by R. Werchota

Aid Harmonization and Alignment (2006) 'Initiatives for Zambia: Country-Level Harmonization' *Aid Harmonization and Alignment* (http://www.aidharmonization.org/ah-cla/ah-browser/index-abridged?rgn_cnt=zm&master=master, accessed September 13, 2006)

Allison, M. (2002) 'Balancing Responsibility for Sanitation' *Social Science and Medicine*, 55 (9), 1539–1551.

AMCOW [African Ministers' Council on Water] (2006) 'CSO and Public Participation in the African Ministers Council on Water (AMCOW) Process: Guidelines for Engagement' AMCOW (http://www.amcow.org/EXCO/CSO&AMCOW.doc, accessed May 8, 2006)

AMCOW-TAC & EU [African Minsters' Council on Water & the European Union] (2003) *The EU Water Initiative: Africa – EU Strategic Partnership on Water Affairs and Sanitation, Outline Strategy and 2004–2005 Work Programme*, Working Document agreed at the Pan-African Implementation and Partnership Conference on Water, Addis Ababa, Ethiopia, 8–13 December 2003, between AMCOW-TAC and EU

Andonova, L.B. & Levy, M.A. (2003) 'Franchising Global Governance: Making Sense of the Johannesburg Type II Partnerships' in O.S. Stokke & O.B.Thommesson eds., *Yearbook of International Cooperation on Environment and Development*, London, Earthscan

Austin, J.E. (2007) 'Sustainability through Partnering: Conceptualising Relationships between Business and NGOs' in P. Glasbergen, F. Biermann & A.J.P. Mol eds., *Partnerships, Governance and Sustainable Development: Reflections on Theory and Practice*, Cheltenham, Edward Elgar

Ayre, G. & Callway, R. (2005) 'Outcomes from the World Summit for Sustainable Development' in G. Ayre & R. Callway eds., *Governance for Sustainable Development: A Foundation for the Future*, London: Earthscan

Backstrand, K. (2006) 'Multi-Stakeholder Partnerships for Sustainable Development: Rethinking Legitimacy, Accountability and Effectiveness' *European Environment* 16, 290–306

Baker, S., Kousis, M., Richardson, D. & Young, S. (1997) 'Introduction: the Theory and Practice of Sustainable Development in EU perspective' in S. Baker, M. Kousis, D. Richardson & S. Young eds., *The Politics of Sustainable Development: Theory, Policy and Practice within the European Union*, London: Routledge

Balloch, S. & Taylor, M. (2001a) 'Introduction' in S. Balloch and M. Taylor eds., *Partnership Working: Policy and Practice*, Bristol: The Policy Press

Balloch, S. & Taylor, M. (2001b) 'Conclusion – can partnerships work?' in S. Balloch and M. Taylor eds., *Partnership Working: Policy and Practice*, Bristol: The Policy Press

Barker, J. (2004) Freshwater Programme Leader and PAWS representative for WWF (World Wide Fund for Nature); Interview: May 4, 2004

Barrow, O. & Jennings, M. (2001) 'Introduction: The Charitable Impulse' in O. Barrow & M. Jennings eds. *The Charitable Impulse: NGOs and Development in East and North-East Africa*, Oxford: James Currey Ltd

Batz, F.J. & van Koppen, P. (2006) *Update Status on African Working Group*, Email sent to all members of the EUWI African Working Group, March 17, 2004

BBC News (2006) 'Direct Aid to Governments "Works"' *BBC News* (http://news.bbc.co.uk/2/hi/europe/4755827.stm, accessed September 13, 2006)

Benn, H. (2005a) 'Foreword' in *Partnerships for Poverty Reduction: Rethinking Conditionality. A UK Policy Paper*, London: Department for International Development, Foreign & Commonwealth Office & HM Treasury

Benn, H. (2005b) 'Keynote Address to Royal Geographic Society Conference, "Water for Life – the UK Input", 22/03/05' *Department for International Development* (http://www.dfid.gov.uk/News/files/Speeches/hilary-water-22mar05.pdf, accessed October 22, 2006)

Benner, T., Ivanova, M.H., Streck, C. & Witte, J.M. (2003) 'Moving the Partnership Agenda to the Next Stage: Key Challenges' in T.Benner, C. Streck & J.M. Witte eds., *Progress or Peril? Networks and Partnerships in Global Environmental Governance: The Post-Johannesburg Agenda*, Berlin/Washington D.C.: Global Public Policy Institute

Benner, T. & Witte, J.M. (2004) 'Everybody's Business: Accountability, Partnerships and the Future of Global Governance' in S. Stern & E. Seligmann eds., *The Partnership Principle: New Forms of Governance in the 21st Century*, London: Archetype

Berntell, A. (2005) 'Foreword' in the UN Millennium Project Task Force on Water and Sanitation Final Report, *Health, Dignity, and Development: What Will it Take?*, New York: SIWI & UN Millennium Project

Biermann, F. (2006) 'Global Governance and the Environment' in M. M. Betsill, K. Hochstetler & D. Stevis eds., *International Environmental Politics*, Basingstoke: Palgrave Macmillan

Biermann, F., Chan, M-s., Mert, A. & Pattberg, P. (2007a) 'Multi-Stakeholder Partnerships for Sustainable Development: Does the Promise Hold?' in P. Glasbergen, F. Biermann & A. P.J. Mol eds., *Partnerships, Governance and Sustainable Development: Reflections on Theory and Practice*, Cheltenham, Edward Elgar

Biermann, F., Mol, A.P.J. & Glasbergen, P. (2007b) 'Conclusion: Partnerships for Sustainability: Reflections on a Future Research Agenda' in P. Glasbergen, F. Biermann & A.P.J. Mol eds., *Partnerships, Governance and Sustainable Development: Reflections on Theory and Practice*, Cheltenham, Edward Elgar

Bovaird, T. (2004) 'Public-Private Partnerships: From Contested

Concepts to Prevalent Practice' *International Review of Administrative Sciences* 70 (2), 199–215

Brewer, B. & Hayllar, M.R. (2005) 'Building Public Trust Through Public-Private Partnerships' *International Review of Administrative Sciences* 71 (3), 475–492

Brinkerhoff, J.M. (2002a) 'Global Public Policy, Partnership, and the Case of the World Commission on Dams' *Public Administration* 62 (3), 324–336

Brinkerhoff, J.M. (2002b) *Partnership for International Development: Rhetoric or Results?*, Boulder, Colorado, Lynne Reiner

Brinkerhoff, J.M. (2007) 'Partnership as a Means to Good Governance: Towards an Evaluation Framework' in P. Glasbergen, F. Biermann & A.P.J. Mol eds., *Partnerships, Governance and Sustainable Development: Reflections on Theory and Practice*, Cheltenham, Edward Elgar

Brinkerhoff, D.W. & Brinkerhoff, J.M. (2004) 'Partnerships between International Donors and Non-Governmental Development Organizations: Opportunities and Constraints' *International Review of Administrative Sciences* 70 (2), 253–270

Bruch, C. & Pendergrass, J. (2003) 'Type II Partnerships, International Law and the Commons' *Georgetown International Environmental Law Review* 15 (4), 855–886

Bruyninckx, H. (2006) 'Sustainable Development: The Institutionalisation of a Contested Policy Concept' in M. M. Betsill, K. Hochstetler & D. Stevis eds., *International Environmental Politics*, Basingstoke: Palgrave Macmillan

Byrne, D. (2001) 'Partnership – Participation – Power: The Meaning of Empowerment in Post-Industrial Society' in S. Balloch and M. Taylor eds., *Partnership Working: Policy and Practice*, Bristol: The Policy Press

Calder, F. (2002) 'Strange New Alliances' *The World Today* (Aug/Sep), 8–9

Caplan, K. (2003) 'The Purist's Partnership: Debunking the Terminology of Partnerships' *Partnership Matters* 1, 31–35

Caplan, K., Jones, D. & Keatman, T. (2001) 'Report on BPD Tri-Sector Workshop' *Business Partnerships for Sustainable Development in Water and Sanitation* (http://www.bpd-waterandsanitation.org/english/docs/tswreport.pdf, accessed February 6, 2003)

Carroll, P. & Steane, P. (2000) 'Public-Private Partnerships: Sectoral Perspectives' in S.P. Osborne ed. *Public-Private Partnerships: Theory and Practice in International Perspective,* London & New York: Routledge

Causer, C. (2005) Law firm staff member and PAWS representative at Kirkpatrick & Lockhart Nicholson Graham LLP; Interview: March 23, 2005

Centre for Democracy and Development (2003) 'CDD's Summary Report on Nigeria's 2003 Elections' *Centre for Democracy and Development* (http://www.cdd.org.uk/resources/elections/nigeria_election_analysis_2003.htm, accessed June 30, 2006).

CGG [Commission on Global Governance] (1995) *Our Global Neighbourhood*, Oxford: Oxford University Press

Chanda, O. (2005) Director of the National Water Supply and Sanitation Council (NWASCO), Zambia; Interview: May 10, 2005

Chauhan, S.K. (1983) *Who Puts the Water in the Taps? Community Participation in Third World Drinking Water, Sanitation and Health*, London: International Institute for Environment and Development

Chisanga, C. (2005) Executive Secretary of the Water and Sanitation Association of Zambia (WASAZA); Interview: May 12, 2005

Chlorine Chemistry Council (2004) 'WAWI – West Africa Water Initiative' *Chlorine Chemistry Council* (http://c3.org/partnerships/wrn/wawi/, accessed October 21, 2004)

CIARIS [Centre for Informatic Apprenticeship and Resources in Social Exclusion] (2004) 'Strategic Principles – Partnership' *Centre for Informatic Apprenticeship and Resources in Social Exclusion* (http://ciaris.ilo.org/english/tos/strprinc/partenar/tx_princ/tx_princp_1.htm, accessed June 4, 2004)

Clark, J. (1995) 'The State, Popular Participation, and the Voluntary Sector' *World Development* 23 (4), 593–601

Clarke, N. (2005) Former PAWS Secretariat member, Department of Environment, Food and Rural Affairs; Interview: June 27, 2005

CNHF [Conrad N. Hilton Foundation] (2002) '$41 Million West Africa Initiative August 20, 2002' *Conrad N. Hilton Foundation* (http://www.hiltonfoundation.org/press_release_details.asp?id=11, accessed September 28, 2004)

CNHF [Conrad N. Hilton Foundation] (2003) 'Bringing Clean Water to West Africa' *Humanitarian Affairs* (http://www.hiltonfoundation.org/pdfs/wawi.pdf, accessed September 28, 2004)

CNHF [Conrad N. Hilton Foundation] (2004a) 'Safe Water Development' *Conrad N. Hilton Foundation* (http://www.hiltonfoundation.org/main.asp?id=56&side=3, accessed September 28, 2004)

CNHF [Conrad N. Hilton Foundation] (2004b) 'WAWI Country Profile: Ghana' *Conrad N. Hilton Foundation* (http://www.hiltonfoundation.org/main.asp?id=79, accessed September 28, 20040

CNHF [Conrad N. Hilton Foundation] (2004c) 'WAWI Country Profile: Mali' *Conrad N. Hilton Foundation* (http://www.hiltonfoundation.org/main.asp?id=80, accessed September 28, 2004)

CNHF [Conrad N. Hilton Foundation] (2004d) 'WAWI Country Profile: Niger' *Conrad N. Hilton Foundation* (http://www.hiltonfoundation.org/main.asp?id=81, accessed September 28, 2004)

Coates, B. (2002) 'More Business as Usual' *Guardian Unlimited* (http://society.guardian.co.uk/societyguardian/story/0,7843,793755,00.html, accessed March 14, 2003)

Colin, J. & Lockwood, H. (2002) 'Research and Surveys Series: Making Innovation Work through Partnerships in Water and Sanitation Projects' *Business Partnerships for Sustainable Development in Water and Sanitation* (http://www.bpd-waterandsanitation.org/english/docs/innovation.pdf, accessed February 6, 2003)

Collier, C. (2000) 'NGOs, the Poor and Local Government' in D. Eade ed., *Development, NGOs and Civil Society*, Oxford: Oxfam

COMPASS [Community Partnerships for Sustainable Resource Management in Malawi] (2006) 'Glossary' *Community Partnerships for Sustainable Resource Management in Malawi* (http://www.compass-malawi.com/cglossary.htm, accessed October 25, 2006)

Craig, D. & Porter, D. (2003) 'Poverty Reduction Strategy Papers: A New Convergence' *World Development* 31 (1), 53–69

Crawford, G. (2003) 'Partnership or Power? Deconstructing the *'Partnership for Governance Reform'* in Indonesia' *Third World Quarterly* 24 (1), 139–159

Cropper, S. (1996) 'Collaborative Working and the Issue of Sustainability' in C. Huxham ed., *Creating Collaborative Advantage*, London: SAGE

Davies, A. (2002) 'Power, Politics and Networks: Shaping Partnerships for Sustainable Communities' *Area* 34 (2), 190–203

Davitt, J. (2004) Finance Service Centre Manager, Yorkshire Water

& Kelda Group Co-ordinator for PAWS, Kelda Group; Interview: March 29, 2004

de Alcántara, C. H. (1998) 'Uses and Abuses of the Concept of Governance' *International Social Science Journal* 50 (1), 103–13

DFID [Department for International Development] (2005) 'Benn Unveils New Approach to Aid' *Department for International Development* (http://www.dfid.gov.uk/news/files/pressreleases/pr-conditionality-paper.asp, accessed March 3, 2005)

DFID [Department for International Development] (2006) 'Sector Wide Approaches (SWAps)' *DFID – Millennium Development Goals: Aid, Trade, Growth and Global Partnership* (http://www.dfid.gov.uk/mdg/aid-effectiveness/swaps.asp, accessed October 9, 2006)

Dodds, F. (2002) 'Multi-Stakeholder Processes and Global Governance' in M. Hemmati *Multi-Stakeholder Processes for Governance and Sustainability*, London: Earthscan

Dossal, A. & Fanzo, M. (2004) 'Partnerships for a Better World: Working Together to Preserve the Planet' *Natural Resources Forum* 28, 333–337

Doyle, B.A. & Corliss, J.M. (2006) *West Africa Water Initiative (WAWI) 2006–2010 Strategic Plan: The Power of Partnership*, WAWI

Edwards, M. & Hulme D. (2000) 'Scaling Up NGO Impact on Development: Learning from Experience' in D. Eade ed. *Development, NGOs and Civil Society*, Oxford: Oxfam

Elliott, L. (2006) 'Minister Advocates Aid Autonomy for Developing States' *Guardian Unlimited* (http://business.guardian.co.uk/story/0,177047800.html, accessed June 12, 2006)

EU [European Union] (2002) 'EU Water Initiative: Water for Life – Health, Livelihoods, Economic Development, Peace and Security. Implementing the Programme of Action of the WSSD' *Johannesburg Summit 2002* (http://www.johannesburgsummit.org/html/sustainable_dev/p2_managing_resources/2508_eu_water_initiative.pdf, accessed January 13, 2002)

EuropeAid (2006) 'ACP-EU Water Facility: In Detail' *EuropeAid* (http://ec.europa.eu/comm/europeaid/projects/water/details_en.htm, accessed June 12, 2006)

European Commission (2001) *European Governance: A White Paper*, Brussels: Commission of the European Communities

EUWI [European Union Water Initiative] (2003a) *2nd Meeting of the EUWI Working Group on Water Supply and Sanitation in Africa,* August 27, 2003, Copenhagen: Ministry of Foreign Affairs (Danida)

EUWI [European Union Water Initiative] (2003b) 'EU Water Initiative: 9th Multi-Stakeholder Forum (MSF) Meeting. 25th November 2003, Brussels. Draft Minutes', attached file in personal email from Les Peacock to Amy Stewart, January 20, 2004

EUWI [European Union Water Initiative] (2003c) *EU Water Initiative: 8th Multi-Stakeholder Forum (MSF) Meeting,* 17–18 June 2003, Athens, Draft Minutes, Athens: EUWI

EUWI [European Union Water Initiative] (2004a) *EU Water Initiative: 10th Multi-Stakeholder Forum (MSF) Meeting,* 16 March 2004, Brussels. Draft Minutes, Brussels: European Commission

EUWI [European Union Water Initiative] (2004b) *Zambia Sector Status Report,* June 2004, Zambia: AMCOW-TAC - Working Group on Water and Sanitation & EU Working Group on Water and Sanitation

EUWI [European Union Water Initiative] (2004c) 'Africa-EU Strategic Water Partnership on Water Affairs and Sanitation: Action Plan Water Supply and Sanitation (Draft Working Document for the WSS Africa Working Group)' *SV: A-EUWI: Meeting of the Water Supply and Sanitation for Africa Working Group*, Brussels 16 June 2004, attached file in personal email from Morgens Bregnbaek (Ministry of Foreign Affairs, Denmark) to Amy Stewart, June 10, 2004

EUWI [European Union Water Initiative] (2004d) 'A Partnership Approach' *EUWI* (http://www.euwi.net/index.php?main=1&sub=2, accessed August 28, 2004)

EUWI [European Union Water Initiative] (2004e) *Zambia Country Dialogue Process Report - Draft Final Report,* June 2004, Zambia: AMCOW-TAC - Working Group on Water and Sanitation & EU Working Group on Water and Sanitation

EUWI [European Union Water Initiative] (2005a) 'EUWI Steering Group Meeting. 11 October 2005, Brussels. Minutes' *Freshwater Action Network* (http://www.freshwateraction.net/library/EUWI%20october%2005%20steering%20group.prn.pdf, accessed May 22, 2006)

EUWI [European Union Water Initiative] (2005b) *Minutes of African WG WSS,* 17th January 2005, Brussels: EUWI

EUWI [European Union Water Initiative] (2005c) *Steering Group Meeting, 23rd May 2005, Draft Minutes*, Brussels: EUWI

EUWI [European Union Water Initiative] (2005d) *Steering Group Discussion Paper 2005/013: Implementation of the Strategy for Development of the EUWI: Key Issues Raised by the MSF*, Brussels: EUWI

EUWI [European Union Water Initiative] (2005e) *Revised Final Draft: Strategy for Development of the EUWI*, Brussels: EUWI

EUWI [European Union Water Initiative] (2005f) *Steering Group Discussion Paper No 2005/015: The Creation of One EUWI Africa Working Group*, Brussels: EUWI

EUWI [European Union Water Initiative] (2005g) *Summary of Joint Meeting EUWI WG WSS with IWRM,* 26 August 2005, Stockholm: EUWI

EUWI [European Union Water Initiative] (2005h) *Multi-Stakeholder Forum (MSF) Meeting,* 25 August 2005, Stockholm. Draft Minutes, Stockholm: EUWI

EUWI [European Union Water Initiative] (2005i) *EU Water Initiative Steering Group Paper No 2005/004: Code of Conduct*, Brussels: EUWI

EUWI [European Union Water Initiative] (2006a) 'The EU Water Initiative' *EUWI* (http://www.euwi.net, accessed May 21, 2006)

EUWI [European Union Water Initiative] (2006b) *Strategy for the African Component of the EU Water Initiative Proposal*, Brussels: EUWI

EUWI [European Union Water Initiative] (2007) *European Union Water Initiative Annual Report 2007 - Draft*, Brussels: EUWI

EUWI [European Union Water Initiative] Africa Working Group (2006a) *Draft AWG Work Plan*, Brussels: EUWI

EUWI [European Union Water Initiative] Africa Working Group (2006b) *Proposed Activities to Secure Funding for the AWG*, Finance Memo presented at the Africa Working Group meeting, 2 May, 2006, Brussels: EUWI

EUWI [European Union Water Initiative] Africa Working Group (2007) *Minutes of meeting of the EUWI AWG (Africa Working Group), August 13, 2007*, Stockholm: EUWI

EUWI [European Union Water Initiative] Secretariat (2006) *EU Perspectives on the Africa-EU Strategic Partnership on Water Affairs and Sanitation within the EU Water Initiative*, Brussels: EUWI

EUWI [European Union Water Initiative] Working Group on Water

Supply and Sanitation in Africa (2004) *Proposal to Accelerate Progress on National Policy Dialogues*, Brussels: EUWI

EUWI [European Union Water Initiative] Working Group on Water Supply and Sanitation in Africa (2005) *Country Level Dialogues Key Documents*, Brussels, EUWI

Evans, B., McMahon, J. & Caplan, K. (2004) 'The Partnership Paperchase: Structuring Partnership Agreements in Water and Sanitation in Low-Income Communities' *Building Partnerships for Development* (http://www.bpd-waterandsanitation.org/english/docs/paperchase.pdf, accessed December 16, 2004)

Eweje, G. (2007) 'Strategic Partnerships between MNEs and Civil Society: the Post-WSSD Perspectives' *Sustainable Development* 15 (10), 15–21

Frade, J. (2004) Representative of the European Investment Bank in EUWI; Interview: April 6, 2004

Friends of the Earth (2002) 'Corporate Accountability & the Johannesburg Earth Summit' *Friends of the Earth* (http://www.foe.org/WSSD/partnerships.html, accessed June 27, 2006)

Gadebe, T. (2006) 'South Africa: Country Reaches MDG on Basic Water' *AllAfrica* (http://allafrica.com/stories/200607070813.html, accessed July 31, 2006)

Gallagher, K.P. (2001) 'Reforming Global Institutions: Overview Essay' in J.M. Harris, T.A. Wise, K.P. Gallagher & N.R. Goodwin eds., *A Survey of Sustainable Development: Social and Economic Dimensions*, Washington: Island Press

Gardiner, R. (2002) 'Governance for Sustainable Development: Outcomes from Johannesburg' *Earth Summit 2002 Building Partnerships for Sustainable Development* (http://www.earthsummit2002/es/issues/governanceoutcomes.pdf, accessed December 11, 2002)

Glasbergen, P. (2007) 'Setting the Scene: the Partnership Paradigm in the Making' in P. Glasbergen, F. Biermann & A.P.J. Mol eds., *Partnerships, Governance and Sustainable Development: Reflections on Theory and Practice*, Cheltenham, Edward Elgar

Global Policy Forum (2005) 'The Global Compact's Next Phase' *Global Policy Forum*

(http://www.globalpolicy.org/reform/business/2005/0906framework.pdf, accessed September 13, 2006)

Global Water Partnership (2008) 'GWP Library – Dublin Statements and Principles' *Global Water Partnership* (http://www.gwpforum.org/servlet/PSP?iNodeID=1345, accessed February 2, 2008)

Gramstad, S. (2003) 'Nigeria: Parliamentary, Presidential and States' Elections 2003' *Norwegian Centre for Human Rights* (http://www.humanrights.uio.no/forskning/publ/nr/2003/11.pdf, accessed June 30, 2006)

Greenhill, R. & Wekiya, I. (2004) *Turning off the Taps: Donor Conditionality and Water Privatisation in Dar es Salaam, Tanzania*, London: ActionAid

Grimm, S. (2004) *ODI European Development Cooperation to 2010 Briefing Paper: Aid disbursement and Effectiveness*, London: ODI

Grindle, M.S. (2004) 'Good Enough Governance: Poverty Reduction and Reform in Developing Countries' *Governance: An International Journal of Policy, Administration, and Institutions* 17 (4), 525–548

Gupta, J. (2002) 'Global Sustainable Development Governance: Institutional Challenges from a Theoretical Perspective' *International Environmental Agreements: Politics, Law and Economics* 2, 361–388

Hailey, J. (2000) 'NGO Partners: The Characteristics of Effective Development Partnerships' in S.P. Osborne ed. *Public-Private Partnerships: Theory and Practice in International Perspective*, London & New York: Routledge

Halcrow (2003) Personal communication to Amy Stewart from a member of the Water and Utilities Division, Halcrow, 10 February, 2003

Hale, T. N. & Mauzerall, D.L. (2004) 'Thinking Globally and Acting Locally: Can the Johannesburg Partnerships Coordinate Action on Sustainable Development?' *Journal of Environment and Development* 13 (3), 220–239

Hall, A.W. (2005) 'Water: Water and Governance' in G. Ayre & R. Callway eds., *Governance for Sustainable Development: A Foundation for the Future*, London: Earthscan

Haque, M.S. (2004) 'Governance Based on Partnership with NGOs: Implications for Development and Empowerment in Rural Bangladesh' *International Review of Administrative Sciences* 70 (2), 271–290

Harris, J.M. & Goodwin, N.R. (2001) 'Volume Introduction' in J.M. Harris, T.A. Wise, K.P. Gallagher & N.R. Goodwin eds., *A Survey of*

Sustainable Development: Social and Economic Dimensions, Washington: Island Press

Hemmati, M. (2002) *Multi-Stakeholder Processes For Governance and Sustainability*, London: Earthscan

Hemmati, M. & Whitfield, R. (2003a) 'Capacity Building for Sustainable Development Partnerships: A Template for Stakeholders, Governments, and Agencies' *Earth Summit 2002 Building Partnerships for Sustainable Development* (http://www.earthsummit2002.org/es/preparations/global/capacity%20building.pdf, accessed May 15, 2003)

Hemmati, M. & Whitfield, R. (2003b) 'Sustainable Development Partnerships in the Follow-up to Johannesburg: Suggestions for Effective Mechanisms at the Regional and International Level' *Earth Summit 2002 Building Partnerships for Sustainable Development* (http://www.earthsummit2002.org/es/preparations/global/partnerships.pdf, accessed May 15, 2003)

HRH the Prince of Orange (2002) 'No Water No Future: A Water Focus for Johannesburg' *Consultative Group on International Agriculture* (http://www.cgiar.org/iwmi/accra2002/NoWaterNoFuture.pdf, accessed March 10, 2003)

Huillet, C. (2004) 'Foreword' in W.R. Lovan, M. Murray & R. Shaffer eds., *Participatory Governance: Planning, Conflict Mediation and Public Decision-Making in Civil Society*, Aldershot: Ashgate

Huxham, C. (1996) 'Introduction: Collaboration and Collaborative Advantage' in C. Huxham ed., *Creating Collaborative Advantage*, London: SAGE

Huxham, C. & Vangen, S. (2000) 'What Makes Partnerships Work?' in S.P. Osborne ed., *Public-Private Partnerships: Theory and Practice in International Perspective*, London & New York: Routledge

IDSUH, IDSRU & CMI [Institute of Development Studies – University of Helsinki, Department of International Development Studies – Roskilde University Centre & Chr. Michelsen Institute] (2002) 'Research Design – Consultation and Empowerment: Governance implications of Participatory Public Policy Formulation in Five Developing Countries. A Collaborative Policy Research Programme' *University of Helsinki – Faculty of Social Sciences* (http://www.valt.helsinki.fi/kmi/english/conf/tpoc/research_plan.htm, accessed September 22, 2004)

Ingram, G.K. (2004) 'Overview' in A. Liebenthal, O.N. Feinstein

& G.K. Ingram eds., *Evaluation and Development: the Partnership Dimension*, New Brunswick, Transaction Publishers

Interviewee 1 (2003) Representative of an NGO partner organization in PAWS and the EUWI: November 11, 2003

Interviewee 2 (2003) Representative of a not-for-profit partner organization in PAWS and the EUWI: November 12, 2003

Interviewee 3 (2003) Representative of an NGO partner organization in PAWS and the EUWI: November 14, 2003

Interviewee 4 (2004) European Commission official – member of the EUWI Secretariat: March 29, 2004

Interviewee 5 (2004) Ex-European Commission official, ex-co-ordinator of the EUWI, and ex-chair of the EUWI Multi-Stakeholder Forum: March 31, 2004

Interviewee 6 (2004) European Commission official – member of EUWI Secretariat: April 1, 2004

Interviewee 7 (2004) European Commission official – member of EUWI Secretariat: March 12, 2004, and April 5, 2004

Interviewee 8 (2004) Representative of a water company partner organization in EUWI: March 17, 2004

Interviewee 9 (2004) Representative of an African NGO organization and member of the African Civil Society Network on Water and Sanitation: April 21, 2004

Interviewee 10 (2004) Independent consultant and NGO representative in the EUWI: April 22, 2004

Interviewee 11 (2004) European Commission official – member of the EUWI Secretariat: November 17, 2004

Interviewee 12 (2005) Representative of an EU Member State in the EUWI: April 4, 2005

Interviewee 13 (2005) Representative of an EU Member State in the EUWI: January 28, 2005

Interviewee 14 (2004) Representative of water company partner organization in PAWS: April 5, 2004

Interviewee 15 (2004) Representative of a WAWI partner organization: November 29, 2004

Interviewee 16 (2005) Representative of Development Cooperation Ireland (DCI) in the EUWI within Zambia: May 16, 2005

Interviewee 17 (2005) Zambian Country Representative for an international NGO and member of the EUWI: May 30, 2005

Interviewee 18 (2005) International donor representative: June 2, 2005

Interviewee 19 (2005) Zambian Ministry of Energy and Water Development (MEWD) official: July 27, 2005

Interviewee 20 (2005) International donor representative: May 31, 2005

Interviewee 21 (2005) Representative of the Mvula Trust: September 16, 2005

Interviewee 22 (2005) Representative of the World Bank and the Water and Sanitation Programme in Zambia: May 12, 2005

Interviewee 23 (2005) Representative of a WAWI partner organization: March 23, 2005

Interviewee 24 (2004) Representative of a WAWI partner organization: November 30, 2004

Interviewee 25 (2004) Representative of a WAWI partner organization: December 1, 2004 and December 2, 2004

Interviewee 26 (2004) Representative of a WAWI partner organization: December 15, 2004

Interviewee 27 (2004) Representative of a WAWI partner organization: December 14, 2004

Interviewee 28 (2005) Representative of a WAWI partner organization: January 28, 2005

Interviewee 29 (2005) Representative of a WAWI partner organization: February 8, 2005

Ivanova, M. (2003) 'Partnerships, International Organizations, and Global Environmental Governance' in T.Benner, C. Streck & J.M. Witte eds., *Progress or Peril? Networks and Partnerships in Global Environmental Governance: The Post-Johannesburg Agenda*, Berlin/ Washington D.C.: Global Public Policy Institute

Jessop, B. (1997) 'Capitalism and its Future: Remarks on Regulation, Government and Governance' *Review of International Political Economy* 4 (3), 561–581

Kadimba-Mwanamwambwa, C. (2005) Strategic Planning and Advocacy Coordinator in the Organizational Learning Unit, Care International, Zambia; Interview: May 20, 2005

Kapotwe, P. (2005) Representative from Oxfam in Zambia; Interview: June 1, 2005

Kara, J. & Quarless, D. (2002) 'Guiding Principles for Partnerships for

Sustainable Development ("Type 2 Outcomes") to be Elaborated on by Interested Parties in the Context of the World Summit on Sustainable Development (WSSD)' *Johannesburg Summit 2002* (http://www.un.org/esa/sustdev/partnerships/guiding_principles7june2002.pdf, accessed April 29, 2006)

Kayizzi-Mugerwa, S. (1998) 'Africa and the Donor Community: From Conditionality to Partnership' *Journal of International Development* 10, 219–225

Khambule, T. (2005) PAWS Co-ordinator for South Africa; Interview: September 16, 2005

Klitgaard, R. (2004) 'Evaluation of, for and through Partnerships' in A. Liebenthal, O.N. Feinstein & G.K. Ingram eds., *Evaluation and Development: the Partnership Dimension*, New Brunswick, Transaction Publishers

Knighton, D. (2002) 'Thirsty World' *The World Today* (Aug/Sept), 13–14

Kooiman, J. (2000) 'Societal Governance: Levels, Modes, and Orders of Social-Political Interaction' in J. Pierre ed. *Debating Governance*, Oxford: Oxford University Press

Küpçü, M. F. (2005) 'Society: Participation and Engagement' in G. Ayre & R. Callaway eds., *Governance for Sustainable Development: A Foundation for the Future*, London: Earthscan

Labre, J. (2004) Representative for Suez in EUWI; Interview: November 3, 2004

Lasker, R.D., Weiss, S. & Miller, R. (2001) 'Partnership Synergy: A Practical Framework for Studying and Strengthening the Collaborative Advantage' *The Milbank Quarterly* 79 (2), 179–205

Le Génissel, M. (2004) Representative for French Department of Natural Resources and the Environment in the EUWI; Interview: November 5, 2004

Legoabe, A. (2005) Contract and Legal Services Manager, Department for Provincial and Local Government, South Africa; Interview: September 9, 2006

Lewis, D. (2000) 'Building "Active" Partnerships in Aid Recipient Countries: Lessons from a Rural Development Project in Bangladesh' in S.P. Osborne ed., *Public-Private Partnerships: Theory and Practice in International Perspective*, London & New York: Routledge

Lister, S. (2000) 'Power in Partnership? An Analysis of an NGO's Relationships with its Partners' *Journal of International Development* 12, 227–239

Lovan, W.R., Murray, M. & Shaffer, R. (2004) 'Participatory Governance in a Changing World' in W.R. Lovan, M. Murray & R. Shaffer eds., *Participatory Governance: Planning, Conflict Mediation and Public Decision-Making in Civil Society*, Aldershot: Ashgate

Lowndes, V. & Skelcher, C. (1998) 'The Dynamics of Multi-Organizational Partnerships: An Analysis of Changing Modes of Governance' *Public Administration* 76, 313–333

Lubinga, F. (2005) Programme Office for the Japanese International Cooperation Agency (JICA); Interview: June 1, 2005

Lupunga, P. (2005) Representative from the Zambian Ministry of Finance, responsible for aid harmonization and the German desk; Interview: May 9, 2005

Mackintosh, M. (1992) 'Partnership: Issues of Policy and Negotiation' *Local Economy* 7 (3), 210–224

Malloch Brown, M. (2004) 'Partnering with the Private Sector in Development' in S. Stern & E. Seligmann eds., *The Partnership Principle: New Forms of Governance in the 21st Century*, London: Archetype

Manz, T. (2007) 'Preface' in J. Martens *Multistakeholder Partnerships: Future Models of Multilateralism?*, Occasional Papers No 29 January 2007, Berlin, Friedrick-Ebert-Stiftung

Margolis, J. (2004) 'Partnerships for Sustainable Development: Building on the Johannesburg Legacy' *U.S. Department of State* (http://www.state.gov/g/oes/rls/rm/2004/34337.htm, accessed October 10, 2004)

Mariotti, J. L. (1996) *The Power of Partnerships*, Cambridge: Blackwell Business

Martens, J. (2007) *Multistakeholder Partnerships: Future Models of Multilateralism?*, Occasional Papers No 29 January 2007, Berlin, Friedrick-Ebert-Stiftung

Matsui, Y. (1995) 'The Road to Sustainable Development: Evolution of the Concept of Development in the UN' in K. Ginther, E. Denters & P.J.I.M. de Waart eds., *Sustainable Development and Good Governance,* Dordrecht: Matrinus Nijhoff

Maxwell, S. & Conway, T. (2000) *Perspectives on Partnership*, Washington, D.C.: World Bank Operations Evaluation Department

Mayo, M. & Taylor, M. (2001) 'Partnership and Power in Community Regeneration' in S. Balloch and M. Taylor eds., *Partnership Working: Policy and Practice*, Bristol: The Policy Press

Mbassi, J.P.E. (2004) *What Could be the Role of Local Authorities in the Africa-EU Water Initiative?,* Stockholm: Presentation given on 17th August 2004 during World Water Week

McDonald, I. (2005) 'Theorising Partnerships: Governance, Communicative Action and Sport Policy' *Journal of Social Policy* 34 (4), 579–600

McEwan, C. (2003) "Bringing Government to the People': Women, Local Governance and Community Participation in South Africa' *Geoforum* 34 (4), 469–481

McLaren, D. (2002) 'Corporate Engagement by "Socially Responsible" Investors: A Practical Paradigm for Stakeholder Governance?' *Ashridge Centre for Business and Society* (http://www.acbas.com/Web/ashacbas.nsf/pdfs/MBAwin/$file/MBAwin02.pdf, accessed June 11, 2003)

McQuaid, R.W. (2000) 'The Theory of Partnership: Why Have Partnerships?' in S.P. Osborne ed. *Public-Private Partnerships: Theory and Practice in International Perspective*, London & New York: Routledge

Meadowcroft, J. (2007) 'Democracy and Accountability: the Challenge for Cross-Sectoral Partnerships' in P. Glasbergen, F. Biermann & A.J.P. Mol eds., *Partnerships, Governance and Sustainable Development: Reflections on Theory and Practice*, Cheltenham, Edward Elgar

Mercer, C. (2003) 'Performing Partnership: Civil Society and the Illusions of Good Governance in Tanzania' *Political Geography* 22 (7), 741–763

Mestrum, F. (2003) 'Poverty Reduction and Sustainable Development' *Environment, Development and Sustainability* 5, 41–61

Mitchell, B. (2005) 'Participatory Partnerships: Engaging and Empowering to Enhance Environmental Management and Quality of Life?' *Social Indicators Research* 71, 123–144

Mitchell, E. (2003) Head of Environment & Corporate Responsibility and PAWS representative for RWE Thames; Interview: November 14 2003

Mohan, G. (2002) 'The Disappointments of Civil Society: The Politics of NGO Intervention in Northern Ghana' *Political Geography* 21, 125–154

Mol, A.P.J. (2007) 'Bringing the Environmental State back in: Partnerships in Perspective' in P. Glasbergen, F. Biermann & A.P.J. Mol eds., *Partnerships, Governance and Sustainable Development: Reflections on Theory and Practice*, Cheltenham, Edward Elgar

Morley, D. (2004) Representative for Freshwater Action Network in EUWI and PAWS; Interview: March 12, 2004

Mosse, D. (2003) 'The Making and Marketing of Participatory Development' in P. Quarless Van Ufford & A. Kumar Giri eds., *A Moral Critique of Development: In Search of Global Responsibilities*, London: Routledge

Mulambo, C. (2005) Senior Water and Sanitation Engineer, Department of Infrastructure and Support Services (DISS), Ministry of Local Government and Housing (MLGH), Zambia; Interview: June 1, 2006

Muller, M. (2002) *The National Water and Sanitation Programme in South Africa: Turning the 'Right to Water' into Reality*, Nairobi, Kenya: Water and Sanitation Programme-Africa Region

Müller-Kraenner, S. (2003) 'Partnerships as an Instrument to Implement the Johannesburg Policy Targets' in T.Benner, C. Streck & J.M. Witte eds., *Progress or Peril? Networks and Partnerships in Global Environmental Governance: The Post-Johannesburg Agenda*, Berlin/Washington D.C.: Global Public Policy Institute

Murphy, D.F. & Bendell, J. (1997) *In the Company of Partners*, Bristol: The Policy Press

Murphy, D.F. & Coleman, G. (2000) 'Thinking Partners: Business, NGOs and the Partnership Concept' in J. Bendell ed., *Terms for Endearment: Business, NGOs and Sustainable Development*, Sheffield: Greenleaf Publishing Limited

Murray, S. (2002) 'Partnership Initiatives Information Sheet: Water for the Poor Signature Initiative' *Johannesburg Summit 2002* (http://www.un.org/esa/sustdev/partnerships/activities_initiate/water_poor.PDF, accessed November 24, 2004)

Mushauri, J. (2006) *Mid-Term Review Workshop Africa-EU Strategic Partnership on Water Affairs and Sanitation,* Imperial Resort Beach Hotel, Entebbe, Uganda. 13–14 February, 2006, Entebbe: EUWI

Mwanza, D.D. (2003) 'Water for Sustainable Development in Africa' *Environment, Development and Sustainability* 5 (1–2), 95–115

Mwanza, M. (2005) Consultant employed by German Technical

Co-operation (GTZ) for establishment of EUWI Zambian country dialogue; Interview: May 5, 2005

Ndegwa, S.N. (1996) *The Two Faces of Civil Society: NGOs and Politics in Africa*, Connecticut: Kumarian Press.

Nel, C. (2005) Head of Technical Services, Zululand District Municipality, South Africa; Interview: September 27, 2005

Nerquaye-Tetteh, B. (2005) 'Message from Bismark Nerquaye-Tetteh, WAWI Coordinator' *Water for the Poor* (http://www.waterforthepoor.org, accessed March 1, 2005)

Nerquaye-Tetteh, B. (2006) 'Foreword' in B.A. Doyle & J.M. Corliss, *West Africa Water Initiative (WAWI) 2006–2010 Strategic Plan: The Power of Partnership*, WAWI

Norad [Norwegian Agency for Development Cooperation] (2004) *From Earmarked Sector Support to General Budget Support- Development Partners' Experience*, Oslo: Norad_

Norris, C. (2005) 'Partnerships for Sustainable Development: the Role of Type II Agreements' in A.C. Kallhauge, G. Sjostedt & E. Corell eds., *Global Challenges: Furthering the Multilateral Process for Sustainable Development*, Sheffield, Greenleaf Publishing

Nowak, S. (2003) Policy Co-ordinator and PAWS representative for Water UK; Interview: November 13, 2003

OECD (2006) *Evaluating the Effectiveness and Efficiency of Partnerships*, ENV/EPOC (2006) 15, 16/9/06, Paris, Organization for Economic Cooperation and Development

Oudyn, R. (2004) Director of Communications and PAWS representative for Biwater; Interview: March 16, 2004

Pallemaerts, M. (2003) 'Is Multilateralism the Future? Sustainable Development or Globalisation as "A Comprehensive Vision of the Future of Humanity"' *Environment, Development and Sustainability* 5, 275–295

Pasha, M. K. & Blaney, D. L. (1998) 'Elusive Paradise: The Promise and Peril of Global Civil Society' *Alternatives* 23 (3), 417–450

PAWS [Partners for Water and Sanitation] (2003a) *Report on the Scoping Study Visit to uThungulu, South Africa,* 19th–25th January, attached file in personal email from Les Peacock to Amy Stewart, March 11, 2003

PAWS [Partners for Water and Sanitation] (2003b) *Report on the Scoping Study Working Group Visit to Matjhabeng, Free State,* 10–14 February 2003, attached file in personal email from Les Peacock to Amy Stewart, March 17, 2003

PAWS [Partners for Water and Sanitation] (2003c) *Report on the Scoping Study Working Group Visit to Nkomazi, Ehlanzeni,* 31 March – 4 April 2003, attached file in personal email from Les Peacock to Amy Stewart, May 21, 2003

PAWS [Partners for Water and Sanitation] (2003d) *Report on the Scoping Study Working Group Visit to Zululand, Kwa-Zulu Natal,* 7–11 April 2003, attached file in personal email from Les Peacock to Amy Stewart, May 21, 2003

PAWS [Partners for Water and Sanitation] (2003e) *Draft Note on the Forum Meeting for Partners for Water and Sanitation,* held on Thursday 8 May 2003, attached file in personal email from Les Peacock to Amy Stewart, May 21, 2003

PAWS [Partners for Water and Sanitation] (2003f) 'The Partners' *Partners for Water and Sanitation* (http://www.partnersforwater.org/, July 23, 2003)

PAWS [Partners for Water and Sanitation] (2003g) *Rule of Engagement for Partners' Participation in Projects in South Africa*, attached file in personal email from Les Peacock to Amy Stewart, July 15, 2003

PAWS [Partners for Water and Sanitation] (2004a) *Proposal for Partners for Water and Sanitation Partnership with Nigeria*, attached file in Les Peacock's personal email to Amy Stewart, February 9, 2003

PAWS [Partners for Water and Sanitation] (2004b) *Partners for Water & Sanitation Annual Report 2002–2003*, London: Department for Environment, Food and Rural Affairs

PAWS [Partners for Water and Sanitation] (2004c) *Final Note on the Partnership Forum Meeting,* held on 6 May 2004 at Water UK, attached file in personal email from Les Peacock to Amy Stewart, June 10, 2004

PAWS [Partners for Water and Sanitation] (2004d) *Note of the Partners for Water and Sanitation Forum Meeting,* 27 October 2004, Ashdown House, London: Department for Environment, Food and Rural Affairs

PAWS [Partners for Water and Sanitation] (2004e) *Report on the Partnership Visit to Uthungulu and Zululand,* 24–28 May 2004, attached file in personal email from Les Peacock to Amy Stewart, June 10, 2004

PAWS [Partners for Water and Sanitation] (2004f) *Report on the Partnership Visit to Nkomazi Local Municipality,* 2–6 February 2004, attached file in personal email from Les Peacock to Amy Stewart, June 10, 2004

PAWS [Partners for Water and Sanitation] (2005a) *Partners for Water and Sanitation Terms of Reference*, London: Department for Environment, Food and Rural Affairs

PAWS [Partners for Water and Sanitation] (2005b) *Draft Note: PAWS Steering Group Meeting,* 25 May 2005, London: Department for Environment, Food and Rural Affairs

PAWS [Partners for Water and Sanitation] (2005c) *Partners for Water and Sanitation Newsletter*, 2 December 2005, London: PAWS

PAWS [Partners for Water and Sanitation] (2005d) *Partners for Water and Sanitation Newsletter*, 11 April 2005, London: PAWS

PAWS [Partners for Water and Sanitation] (2005e) *Partners for Water and Sanitation. Water and Sanitation for Africa: Our Performance in 2004, Our Plans for 2005.* Annual Report 2004/2005, London: Department for Environment, Food and Rural Affairs

PAWS [Partners for Water and Sanitation] (2006a) 'About the Partnership – Key Objectives' *Partners for Water and Sanitation* (http://www.partnersforwater.org/about/objectives/, accessed April 26, 2006)

PAWS [Partners for Water and Sanitation] (2006b) 'Partners Directory' *Partners for Water and Sanitation* (http://www.partnersforwater.org/partners/uk/, accessed April 28, 2006)

PAWS [Partners for Water and Sanitation] (2006c) 'Partners for Water and Sanitation' *UN Division for Sustainable Development: Partnerships for Sustainable Development – CSD Partnership Database* (http://webapps01.un.org/dsd/partnerships/public/partnerships/92.html, accessed September 12, 2006)

PAWS [Partners for Water and Sanitation] (2006d) *Confluence: The Newsletter for Partners for Water and Sanitation,* May 2006, London: PAWS

PAWS [Partners for Water and Sanitation] (2006e) 'Where We Work' *Partners for Water and Sanitation* (http:www.partnersforwater.org/countries/where/, accessed July 6, 2006)

PAWS [Partners for Water and Sanitation] (2006f) 'What We Do' *Partners for Water and Sanitation* (http:www.partnersforwater.org/whatwedo/, accessed July 6, 2006)

PAWS [Partners for Water and Sanitation] (2006g) 'About the Partnership' *Partners for Water and Sanitation* (http:www.partnersforwater.org/about/, accessed July 6, 2006)

PAWS [Partners for Water and Sanitation] (2006h) 'How We Work' *Partners for Water and Sanitation* (http:www.partnersforwater.org/whatwedo/how we work/, accessed July 6, 2006)

PAWS [Partners for Water and Sanitation] (2006i) 'PAWS Newsletter July/August 2006' *Partners for Water and Sanitation* (http://www.partnersforwater.org/news/current/, accessed August 17, 2006)

PAWS [Partners for Water and Sanitation] (2006j) *Partners for Water and Sanitation Annual Report 2005–06*, London: PAWS

PAWS [Partners for Water and Sanitation] (2006k) 'Newsletter March 2006' *Partners for Water and Sanitation* (http://www.partnersforwater.org/downloads/march06.pdf, accessed January 26, 2008)

PAWS [Partners for Water and Sanitation] (2007) 'Confluence April 2007' *Partners for Water and Sanitation* (http://www.partnersforwater.org/downloads/apr2007.pdf, accessed January 26, 2008)

PAWS [Partners for Water and Sanitation] Secretariat (2004) Personal communication to Amy Stewart from a member of the PAWS Secretariat, September 7, 2004

Peacock, L. (2003) Former Head of PAWS Secretariat, Department of Environment, Food and Rural Affairs; Interview: November 10, 2003

Pearce, J. (2000) 'Development, NGOs, and Civil Society: the Debate and its Future' in D. Eade ed. *Development, NGOs and Civil Society*, Oxford: Oxfam

Phiri, Z. (2005) Former Programme Manager of Zambia's Water Resources Action Programme (WRAP), and currently Project Manager of the Zambezi River Action Plan Project; Interview: May 31, 2005

Pierre, J. (2000a) 'Introduction: Understanding Governance' in J. Pierre ed., *Debating Governance*, Oxford: Oxford University Press

Pierre, J. (2000b) 'Conclusions: Governance beyond State and Strength' in J. Pierre ed., *Debating Governance*, Oxford: Oxford University Press

Pinto, M.C.W. (1995) 'Reflections on the Term Sustainable Development and its Institutional Implications' in K. Ginther, E. Denters & P.J.I.M. de Waart eds., *Sustainable Development and Good Governance*, Dordrecht: Matrinus Nijhoff

Poncelet, E.C. (2001) 'Personal Transformation in Multistakeholder Environmental Partnerships' *Policy Sciences* 34, 273–301

Poncelet E.C. (2004) *Partnering for the Environment: Multistakeholder Collaboration in a Changing World*, Oxford: Rowman & Littlefield

Pratt, N. (2004) 'Bringing Politics Back in: Examining the Link Between Globalization and Democratization' *Review of International Political Economy* 11 (2), 311–336

Quibell, G. (2005) Technical Support Consultant, Department for International Development Water Resources Project, South Africa; Interview: September 12, 2005

Reid, D. (1995) *Sustainable Development: an Introductory Guide*, London: Earthscan

Rhodes, R.A.W (2000) 'Governance and Public Administration' in J. Pierre ed., *Debating Governance*, Oxford: Oxford University Press

Richardson, D. (1997) 'The Politics of Sustainable Development' in S. Baker, M. Kousis, D. Richardson & S. Young eds., *The Politics of Sustainable Development: Theory, Policy and Practice within the European Union*, London: Routledge

Robinson, D. (1999) 'Partnerships – Practice and Theory' in D. Robinson ed., *Partnership – from Practice to Theory*, Wellington: Institute of Policy Studies

Roe, A. (2005) 'Message from the PAWS Chair' *Partners for Water and Sanitation Newsletter, 2 December 2005,* London: PAWS

Rosenau, J.N. (2000) 'Change, Complexity, and Governance in Globalizing Space' in J. Pierre ed. *Debating Governance*, Oxford: Oxford University Press

Rouse, M. (2003) President, Chairman, and PAWS representative for the International Water Association; Interview: November 11, 2003

Rowe, M. & Devanney, C. (2003) 'Partnership and the Governance of Regeneration' *Critical Social Policy* 23 (3), 375–397

Runciman, D. (2003) 'Partnering the State' *Partnership Matters* 1, 12–15

Sachs, J.D. (2004) 'Getting Priorities Right – An Interview' in S. Stern & E. Seligmann eds., *The Partnership Principle: New Forms of Governance in the 21st Century*, London: Archetype

SALGA [South African Local Government Association] (2006) 'About SALGA' *South African Local Government Association* (http://www.salga.net/home.asp?pid=770, accessed July 4, 2006)

SANGOCO [South African NGO Coalition] (2005) 'South Africa Millennium Development Goals Country Report' *SANGOCO* (http://www.sangoco.org.za/index.php?option=content&task=view&id=188&Itemid=, accessed July 31, 2006)

Sarker, A.E. (2005) 'New Public Management, Service Provision and Non-Governmental Organizations in Bangladesh' *Public Organization Review: A Global Journal* 5, 249–271

Saywell, D. (2004) Programme Manager, and representative in the EUWI for the Water Supply and Sanitation Collaborative Council; Interview: April 16, 2004

Schild, T. (2004) *Planning/Working Session Outcomes of the 5th Ordinary Session of AMCOW*, 20–21 December 2004, Nairobi: EUWI internal Communication and Information System

Schipulle, H. P. (2003) 'Lessons Learned from Multi-Stakeholder Partnerships Supported by German Development Cooperation' in T.Benner, C. Streck & J.M. Witte eds., *Progress or Peril? Networks and Partnerships in Global Environmental Governance: The Post-Johannesburg Agenda*, Berlin/Washington D.C.: Global Public Policy Institute

Secretariat of the Cardoso Panel (2004) 'Workshop Summary: Multi-stakeholder Workshop on Partnership and UN-Civil Society Relationships' *United Nations* (http://www.un.org/reform.pocantico_final.doc, accessed April 11, 2006)

Seetal, A. (2005) Director of Water Allocation, Department of Water Affairs and Forestry, South Africa; Interview: September 12, 2005

Selsky, J.W. & Parker, B. (2005) 'Cross-Sector Partnerships to Address Social Issues: Challenges to Theory and Practice' *Journal of Management* 31 (6), 849–873

Seyfang, G. & Jordan, A. (2002) 'The Johannesburg Summit and Sustainable Development: How Effective are Environmental Mega-conferences?' in O. Schram Stokke and O.B. Thommessen eds., *Yearbook of International Co-operation on Environment and Development 2002/2003* London: Earthscan

Shah, A. (2006) 'The US and Foreign Aid Assistance' *Global Issues – Sustainable Development* (http://www.globalissues.org/TradeRelated/Debt/USAid.asp, accessed October 10, 2006)

Sievers, P. (2005) Programme Co-ordinator for the Water Sector in Zambia, and representative in the EUWI for the Danish International Development Agency (DANIDA); Interview: May 5, 2005

Simpungwe, E., Waalewijn, P. & Raven, B. (2007) 'Multi-Stakeholder Dissonance in the South African Water Arena' in J. Warner ed., *Multi-Stakeholder Platforms for Integrated Water Management*, Aldershot: Ashgate

Steets, J. (2004) 'Accountability, Ignored by Partnerships at Own Peril' *Building Partnerships at Own Peril* (http://bdp-waterandsanitation.org/english/newsfeature.asp, accessed October 5, 2004)

Stern, E. (2004) 'Evaluating Partnerships' in A. Liebenthal, O.N. Feinstein & G.K. Ingram eds., *Evaluation and Development: the Partnership Dimension*, New Brunswick, Transaction Publishers

Stigson, B. (2003) 'A Business Perspective on Partnerships' in T.Benner, C. Streck & J.M. Witte eds., *Progress or Peril? Networks and Partnerships in Global Environmental Governance: The Post-Johannesburg Agenda*, Berlin/Washington D.C.: Global Public Policy Institute

Stigson, B. (2004) Remarks made as a panellist in the high-level segment, *Releasing the Energy of Entrepreneurs & Partnerships*, New York: CSD-12, April 29 2004

Stiles, J. & Williams, P. (2003) 'The Impact of Intent on Public/Private Sector Partnership' *The Seventh International Conference on Global Business and Economic Development* (http://blake.montclair.edu/~cibconf/conference/, accessed January 1, 2006)

Teisman, G.R. & Klijn, E. (2002) 'Partnership Arrangements: Governmental Rhetoric or Governance Scheme?' *Public Administration Review* 62 (2), 197–205

Tennyson, R. (2003) 'The Partnering Toolbook' *The International Business Leaders Forum* (http://www.iblf.org/csr/csrwebassist.nsf/550d4b46b29f68a68525686600 81f938/80256adc002b820480256e46005307c5/$FILE/PartneringToolbook.pdf, accessed March 8, 2004)

Tolentino, A.S. (1995) 'Good Governance Through Popular Participation in Sustainable Development' in K. Ginther, E. Denters & P.J.I.M. de

Waart eds., *Sustainable Development and Good Governance*, Dordrecht: Matrinus Nijhoff

Torjman, S. (1999) 'Partnership: Building Block for a Caring Society' in D. Robinson ed., *Partnership – from Practice to Theory*, Wellington: Institute of Policy Studies

Turner, P. (2005) Secondee to the PAWS Secretariat, Northumbrian Water; Interview: June 27, 2005

Twelvetrees, C. (2005): PAWS representative for the Environment Agency; Interview; March 23, 2005

UN (2002) 'Johannesburg Declaration on Sustainable Development' *UN Department of Economic and Social Affairs Division for Sustainable Development* (http://www.un.org/esa/sustdev/documents/WSSD_POI_PD/English/POI_PD.htm, accessed April 5, 2006)

UN (2004) 'Annan Creates New Advisory Board on Water and Sanitation' *UN News Centre* (http://www.un.org/apps/news/story.asp?NewsID=10149, accessed March 27, 2004)

UN Global Compact (2006a) 'What is the Global Compact' *United Nations Global Compact* (http://www.unglobalcompact.org/AboutTheGC/index.html, accessed September 13, 2006)

UN Global Compact (2006b) 'Global Compact Integrity Measures' *United Nations Global Compact* (http://www.unglobalcompact.org/AboutTheGC/integrity.html, accessed September 13, 2006)

UNCSD [United Nations Commission on Sustainable Development] Secretariat (2005) 'Highlights of the CSD-13 Partnership Fair activities' *United Nations Department for Economic and Social Affairs – Division for Sustainable Development* (http://www.un.org/esa/sustdev/csd/csd13/partnerships_fair.htm, accessed April 18, 2006)

UNDESA [United Nations Department of Economic and Social Affairs] (2002a) 'The Johannesburg Summit Test: What Will Change?' *Johannesburg Summit 2002* (http://www.johannesburgsummit.org/html/whats_new/feature_story41.html, accessed December 4, 2002)

UNDESA [United Nations Department of Economic and Social Affairs] (2002b) 'World Summit on Sustainable Development Plan of Implementation' *Johannesburg Summit 2002* (http://www.johannesburgsummit.org/html/documents/summit_docs/2309_planfinal.htm, accessed January 13, 2003)

UNDESA [United Nations Department of Economic and Social Affairs]

(2002c) 'Major Groups and the World Summit on Sustainable Development' *Johannesburg Summit 2002* (http://www.johannesburg-summit.org/html/major_groups/majorgrpandsummit.html, accessed April 29, 2006)

UNDESA [United Nations Department of Economic and Social Affairs] (2005) 'CSD-13 Partnerships Fair – April 2005, CSD Secretariat's Note: Highlights from the "Partnerships in Practice" Interactive Discussion Sessions' *United Nations Department of Economic and Social Affairs*_(http://www.un.org/esa/sustdev/csd/csd13/PF_highlights.pdf#search=%22csd%2013%20partnership's%20fair%20discussion%20session%22, accessed May 13, 2006)

UNDP [United Nations Development Programme] (1997) *Governance for Sustainable Human Development*, New York: UNDP

UNDP [United Nations Development Programme] (2006) *Human Development Report – Beyond Scarcity: Power, Poverty and the Global Water Crisis*, New York: UNDP

UNDSD [United Nations Division for Sustainable Development] (2003) 'CSD-11 Decision on Partnerships' *UN Department of Economic and Social Affairs – Division for Sustainable Development* (http://www.un.org/esa/sustdev/partnerships/csd11_partnerships_decision.htm, accessed April 29, 2006)

UNDSD [United Nations Division for Sustainable Development] (2005) 'Partnerships for Sustainable Development – Update' *UN Department of Economic and Social Affairs – Division for Sustainable Development* (http://www.un.org/esa/sustdev/csd/csd13/documents/bground_1.pdf, accessed March 22, 2006)

UNDSD [United Nations Division for Sustainable Development] (2006a) 'About CSD' *UN Department of Economic and Social Affairs – Division for Sustainable Development* (http://www.un.org/esa/sustdev/csd/aboutCsd.htm, accessed March 22, 2006)

UNDSD [United Nations Division for Sustainable Development] (2006b) 'Commission on Sustainable Development – CSD-12 Partnerships Fair' *UN Department of Economic and Social Affairs – Division for Sustainable Development* (http://www.un.org/esa/sustdev/csd/csd12/CSD12_partnerships_fair.htm, accessed October 18, 2006)

UNDSD [United Nations Division for Sustainable Development] (2008a) 'Partnerships for Sustainable Development – CSD Partnerships

Database' *UN Department of Economic and Social Affairs – Division for Sustainable Development* (http://webapps01.un.org/dsd/partnerships/public/welcome.do, accessed January 26, 2008)

UNDSD [United Nations Division for Sustainable Development] (2008b) 'Partnerships for Sustainable Development – CSD Partnerships Database: Partnership Statistics' *UN Department of Economic and Social Affairs – Division for Sustainable Development* (http://webapps01.un.org/dsd/partnerships/public/statisticsResults.do, accessed January 26, 2008)

UNDSD [United Nations Division for Sustainable Development] (2008c) 'Partnerships for Sustainable Development – Partnerships e-Forum' *UN Department of Economic and Social Affairs – Division for Sustainable Development* (http://www.un.org/esa/sustdev/partnerships/eforum/index.htm, accessed February 2, 2008)

UNESC [United Nations Economic and Social Council] (2004) 'Commission on Sustainable Development Twelfth Session: Partnerships for Sustainable Development - Report of the Secretary-General' *United Nations Division for Sustainable Development* (http://daccessdds.un.org/doc/UNDOC/GEN/N04/237/63/PDF/N0423763.pdf?OpenElement, accessed April 29, 2006)

UNESC [United Nations Economic and Social Council] (2006a) 'Commission on Sustainable Development Fourteenth Session: Partnerships for Sustainable Development - Report of the Secretary General' *United Nations Division for Sustainable Development* (http://daccessdds.un.org/doc/UNDOC/GEN/N06/259/15/PDF/N0625915.pdf?OpenElement, accessed November 8, 2006)

UNESC [United Nations Economic and Social Council] (2006b) 'Commission on Sustainable Development Report on the Fourteenth Session' *United Nations Economic Commission for Europe* (http://www.unece.org/ie/se/pdfs/comm15/E.2006.29.SUPP_e.pdf, accessed November 8, 2006)

UNMPTF-WS [United Nations Millennium Project Task Force on Water and Sanitation] (2005) *Health, Dignity, and Development: What Will It Take?,* New York: SIWI & United Nations Millennium Project

USAID [United States Agency for International Development] (2002a) 'Press Release August 20, 2002: $41 Million Public-Private Partnership to Provide Clean Water in West Africa' *USAID* (http://

www.usaid.gov/press/releases/2002/pr020820.html, accessed October 21, 2004)

USAID [United States Agency for International Development] (2002b) 'Press Release August 29, 2002: Final Draft, $41 Million Public-Private Partnership to Provide Clean Water in West Africa' *Water for the Poor* (http://www.waterforthepoor.org/pdfs/usaid_press_release_wawi_wssd.pdf, accessed October 21, 2004)

USAID [United States Agency for International Development] (2004a) 'Water for the Poor Signature Initiative: Fact Sheet, January 29, 2004' *USAID: Sustainable Development Partnerships* (http://www.sdp.gov/initiative/wfp/28307.htm, accessed February 5, 2004)

USAID [United States Agency for International Development] (2004b) '2004 USAID Summer Seminar Series - General Budget Support: A New Approach from Other Donors' *USAID Policy Program Coordination* (http://www.usaid/policy/cdie/7-27-04.htm, accessed September 13, 2006)

USAID [United States Agency for International Development] (2005) *General Budget Support: Key Findings of Five USAID Studies,* PPC Evaluation Paper No. 7, Silver Spring, Maryland: USAID's Development Experience Clearinghouse

USAID [United States Agency for International Development] (2006) 'Partnership Meetings – WAWI' *Mexico 2006 4th World Water Forum* (http://www.worldwaterforum4.org.mx/home/detalle_part.asp?cve_partner=5&=lan, accessed January 1, 2006)

USAID [United States Agency for International Development] Water Team (2004a) *Water for the Poor Initiative: Recent Results,* UN Commission on Sustainable Development, 12th Session - New York: USAID

USAID [United States Agency for International Development] Water Team (2004b) *The West Africa Water Initiative*, UN Commission on Sustainable Development, 12th Session - New York: USAID

Utting, P. (2002) 'The Global Compact and Civil Society: Averting a Collision Course' *United Nations Research Institute for Social Development* (http://www.unrisd.org/unrisd/website/newsview.nsf/(httpNews)/E58C1A77E37FA9F0C1256C7E00490304?OpenDocument, accessed October 17, 2006)

Utting, P. (2003) 'The Global Compact: Why All the Fuss?' *United Nations Research Institute for Social Development* (http://www.unrisd.

org/80256B3C005BE6B5/(search)/BC65635F3740379BC1256D010036BA3E?Opendocument&highlight=2,global,compact&fromsearch=yes&query=global+compact, accessed September 13, 2006)

UWASNET [Ugandan Water and Sanitation NGO Network] (2006) 'About Us' *UWASNET* (http://www.uwasnet.org/about.html, accessed June 29, 2006)

van Edig, A. (2004) Representative for German Federal Ministry of Economic Cooperation and Development in the EUWI; Interview: November 11, 2004

van Huijstee, M.M., Francken, M. & Leroy, P. (2007) 'Partnerships for Sustainable Development: A Review of Current Literature' *Environmental Sciences* 4 (2), 75–89

Waller, R. (2003) Representative of Mouchel Parkman in PAWS; Interview: November 13, 2003

Walshe, M., Liebaert, A., Detoc, S., De Raeve, R., Bardram, E., Vergos, Z. & Nauen, C. (2004) *EU Water Initiative: Organisational Framework and Modalities*, Final Version (9 July 2004), attached file in email from Jutta Neusser to all EUWI mailing list members, July 9, 2004

Water for the Poor (2002a) 'Working Together for a Sustainable World' *Water for the Poor* (http://www.waterforthepoor.org/, accessed March 19, 2004)

Water for the Poor (2002b) 'West Africa Water Initiative (WAWI)' *Water for the Poor* (http://www.waterforthepoor.org/initiatives/wawi/wawi.htm, accessed March 19, 2004)

Water for the Poor (2002c) 'WAWI Brochure' *Water for the Poor* (http://www.waterforthepoor.org, 10, accessed October, 2004)

Water for the Poor (2005a) 'West Africa Water Initiative (WAWI)' *Water for the Poor* (http://www.waterforthepoor.org/initiatives/wawi/wawi.htm, accessed 1 March, 2005)

Water for the Poor (2005b) 'Ghana Activities' *Water for the Poor* (http://www.waterforthepoor.org/initiatives/wawi/ghana.htm, accessed 1 March, 2005)

Water for the Poor (2005c) 'Niger Activities' *Water for the Poor* (http://www.waterforthepoor.org/initiatives/wawi/niger.htm, accessed 1 March, 2005)

Water for the Poor (2005d) 'Mali Activities' *Water for the Poor* (http://www.waterforthepoor.org/initiatives/wawi/mali.htm, accessed 1 March, 2005)

WaterAid (2001) *WaterAid Zambia Country Strategy 2001–2005*, London: WaterAid

WaterAid & Tearfund (2005) *An Empty Glass: The EU Water Initiative's Contribution to the Water and Sanitation Millennium Targets, Discussion Paper*, London: WaterAid and Tearfund

WAWI [West Africa Water Initiative] (2006) 'West Africa Water Initiative (WAWI)' *UN Department of Economic and Social Affairs – Division for Sustainable Development: Partnerships for Sustainable Development – CSD Partnership Database* (http://webapps01.un.org/dsd/partnerships/public/partnerships/1550.html, accessed April 20, 2006)

WCED [World Commission on Environment and Development] (1987) *Our Common Future*, Oxford: Oxford University Press

Weiss, T.G. (2000) 'Governance, Good Governance and Global Challenges: Conceptual and Actual Challenges' *Third World Quarterly* 21 (5), 795–814

Werchota, R. (2005) Programme Manager for German Technical Cooperation (GTZ), the co-ordinating body for the EUWI in Zambia; Interview: May 6, 2005

Weston, D. (2005) Department of Water Affairs and Forestry, South Africa; Interview: September 14, 2005

White, S.C. (2000) 'Depoliticising Development: The Uses and Abuses of Participation' in D. Eade ed., *Development, NGOs and Civil Society*, Oxford: Oxfam

WHO/UNICEF JMP [World Health Organization / United Nations Children's Fund Joint Monitoring Programme for Water Supply and Sanitation] (2004) 'Meeting the MDG Drinking Water & Sanitation Target: A Mid-Term Assessment of Progress' *World Health Organization* (http://www.who.int/water_sanitation_health/monitoring/en/jmp04.pdf, accessed September 20, 2004)

Wilson, A. & Charlton, K. (1997) *Making Partnerships Work: A Practical Guide for the Public, Private, Voluntary and Community Sectors*, York: Joseph Rowntree Foundation

Wise, T.A. (2001) 'Economics of Sustainability: The Social Dimension' in J.M. Harris, T.A. Wise, K.P. Gallagher & N.R. Goodwin eds.,

A Survey of Sustainable Development: Social and Economic Dimensions, Washington: Island Press

Witte, J.M. & Streck, C. (2003) 'Introduction: Progress or Peril? Partnerships and Networks in Global Environmental Governance' in T.Benner, C. Streck & J.M. Witte eds., *Progress or Peril? Networks and Partnerships in Global Environmental Governance: The Post-Johannesburg Agenda*, Berlin/Washington D.C.: Global Public Policy Institute

Witte, J.M., Streck, C. & Benner, T. (2003) 'The Road from Johannesburg: What Future for Partnerships in Global Environmental Governance?' in T.Benner, C. Streck & J.M. Witte eds., *Progress or Peril? Networks and Partnerships in Global Environmental Governanc: The Post-Johannesburg Agenda*, Berlin/Washington D.C.: Global Public Policy Institute

World Vision (2002) 'Background on the West Africa Water Initiative' *World Vision United States* (http://www.worldvision.org/worldvision/pr.nsf/0, accessed October 21, 2004)

Zadek, S. (2003) 'Partnership Futures' *Partnership Matters* 1, 8–11

INDEX

Note: This Index does not include the following items or their derivatives, because there were too many references to them in the text: Africa; government; NGO; participation; partnership; stakeholder.